AF386996

Lecture Notes in Civil Engineering

Volume 851

Lecture Notes in Civil Engineering (LNCE) publishes the latest developments in Civil Engineering—quickly, informally and in top quality. Though original research reported in proceedings and post-proceedings represents the core of LNCE, edited volumes of exceptionally high quality and interest may also be considered for publication. Volumes published in LNCE embrace all aspects and subfields of, as well as new challenges in, Civil Engineering. Topics in the series include:

- Construction and Structural Mechanics
- Building Materials
- Concrete, Steel and Timber Structures
- Geotechnical Engineering
- Earthquake Engineering
- Coastal Engineering
- Ocean and Offshore Engineering; Ships and Floating Structures
- Hydraulics, Hydrology and Water Resources Engineering
- Environmental Engineering and Sustainability
- Structural Health and Monitoring
- Surveying and Geographical Information Systems
- Indoor Environments
- Transportation and Traffic
- Risk Analysis
- Safety and Security

To submit a proposal or request further information, please contact the appropriate Springer Editor:

- Pierpaolo Riva at pierpaolo.riva@springer.com (Europe and Americas);
- Swati Meherishi at swati.meherishi@springer.com (Asia—except China, Australia, and New Zealand);
- Wayne Hu at wayne.hu@springer.com (China).

All books in the series now indexed by Scopus and EI Compendex database!

Marwan Alzaylaie · Saeed Alzahrani ·
Deepankar Choudhury · Robert Liang · Jie Han
Editors

Geotechnical Innovation

Select Proceedings of the 2nd International Geotechnical Innovation Conference (IGIC 2025)

Springer

Editors
Marwan Alzaylaie 
Dubai Development Authority
Dubai, United Arab Emirates

Saeed Alzahrani
Imam Abdulrahman Bin Faisal University
Dammam, Saudi Arabia

Deepankar Choudhury
Indian Institute of Technology (IIT)
Mumbai, Maharashtra, India

Robert Liang
University of Dayton
Dayton, OH, USA

Jie Han
University of Kansas
Lawrence, KS, USA

ISSN 2366-2557   ISSN 2366-2565 (electronic)
Lecture Notes in Civil Engineering
ISBN 978-981-95-9214-2   ISBN 978-981-95-9215-9 (eBook)
https://doi.org/10.1007/978-981-95-9215-9

Contents

Numerical Study on the Dynamic Behavior of Connected and Non-connected CPRF Systems 1
Ashutosh Kumar, Tanmoy Das, and Deepankar Choudhury

Failure Analysis of a Large-Scale Hoarding in Mumbai: Lessons in Wind Load Design and Foundation Safety 7
Pranoy Debnath, Arpita Ray, and Deepankar Choudhury

Micropiles in Urban Infrastructure: A Case Study of Skywalk Construction Connecting Riyadh Airport Terminal with Metro Line 15
Haytham T. S. Hamdan and Hameeduddin Mohammed

Recycling of Crushed Limestone and Tires Rubber Waste for Use as a Geotechnical Material, Initial Assessment 29
Saad Alsabr, Badr Almutairi, Mohamed Farid Abbas, and Sultan Almuaythir

Bored Pile Capacity Prediction Based on CPT Data Using Machine Learning Models 37
Abdulhakim Mawas and Omar Hamza

Evaluation of GCC Desert Sands as Potential Martian Regolith Simulants 57
W. Jrad and S. Alasadi

Numerical Study on the Dynamic Behavior of Connected and Non-connected CPRF Systems

Ashutosh Kumar⬤, Tanmoy Das⬤, and Deepankar Choudhury⬤

Abstract This paper presents a numerical study on the static and dynamic behavior of Combined Piled Raft Foundations (CPRF) with connecting and non-connecting configurations. Using finite element modeling, the influence of cushion layer thickness on stress distribution and seismic response is examined. The analysis focuses on key parameters such as bending moment and acceleration amplification under both static and dynamic loading conditions. Results indicate that CPRFs with thicker cushion layers exhibit more uniform stress distribution and improved seismic performance. In particular, non-connecting configurations benefit significantly from increased cushion thickness, which helps mitigate seismic wave transmission and reduce acceleration amplification in the superstructure. The findings contribute to the broader understanding of soil–structure interaction and provide practical guidance for optimizing foundation performance in urban infrastructure.

Keywords Combined Pile Raft Foundation (CPRF) · Seismic conditions · Finite element · Cushion layer · Amplification

A. Kumar
Department of Atomic Energy, Nuclear Power Corporation of India Limited (NPCIL), Mumbai, India

T. Das (✉)
Postdoctoral Research Associate, School of Civil and Environmental Engineering, University of Technology Sydney, 15 Broadway, Ultimo, NSW 2007, Australia
e-mail: tanmoy.das.biet@gmail.com

D. Choudhury
Department of Civil Engineering, Indian Institute of Technology Bombay, IIT Bombay, Powai, Mumbai, India
e-mail: dc@civil.iitb.ac.in

M. Alzaylaie et al. (eds.), *Geotechnical Innovation*, Lecture Notes in Civil Engineering 851, https://doi.org/10.1007/978-981-95-9215-9_1

1

1 Introduction

The increasing demand for high-rise structures on soft soils has highlighted the limitations of conventional shallow foundations, particularly in controlling excessive and differential settlements [1]. Historical failures, such as the settlement issues faced by the Frankfurt Buro Center in Germany, emphasized the need for more reliable foundation systems. In response, the Combined Piled Raft Foundation (CPRF) concept was developed, which integrates a limited number of piles with a raft to improve load sharing and reduce overall settlement.

Despite their benefits, the behavior of CPRFs, especially the role of pile-raft connection and cushion layer thickness under dynamic loading, remains an area requiring further investigation [2–4]. This study presents a numerical analysis of connecting and non-connecting CPRFs, focusing especially on the seismic performance of CPRF systems in loose sandy soil to evaluate the influence of cushion layers on stress distribution and structural response.

2 Methodology

The finite element models of the CPRF and the Non-Connecting Piled Raft Foundation (NCPRF) are developed in the finite element (FE) based computer program Midas GTS NX 2020 (v1.1) to simulate their behavior under static and dynamic loads. The soil domain measures 30×30 m horizontally and extends 16 m in depth, as shown in Fig. 1a. These dimensions are selected based on established guidelines recommending a minimum of three times the foundation width in the horizontal direction and at least twice the foundation depth vertically to replicate realistic stress distribution and deformation patterns.

The CPRF foundation system studied by Azizkandi et al. [5] was used to validate the present FE model, which includes a raft of 5×5 m in plan with a thickness of 1 m

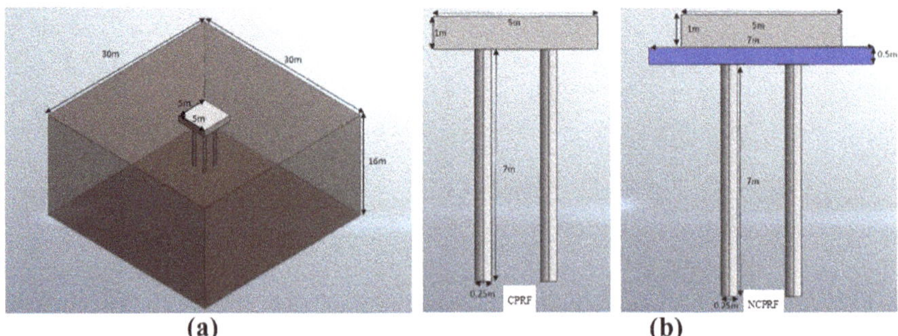

(a)   (b)

Fig. 1 Finite element model: **a** 3D soil domain with CRPF system; **b** configuration of CRPF and NCPRF systems

Table 1 Properties of soil used for numerical study

Parameters	Magnitude
Unit weight	18 kN/m^3
Poisson's ratio	0.24
Secant stiffness in triaxial test	32 GPa
Tangential stiffness	22.4 GPa
Unloading reloading stiffness	96 GPa
Power of stress level dependency	0.5
Friction angle	36^0
Failure ratio	0.9

Table 2 Properties of the pile, raft, and cushion layer

Parameter	Raft	Pile	Cushion layer
Unit weight (kN/m^3)	27.0	27.0	20
Young's modulus (GPa)	210	210	68
Poisson's ratio	0.3	0.2	0.3
Friction angle	–	–	36°

and a group of piles of 7 m in length and 0.25 m in diameter, designed to penetrate into deeper soil layers for enhanced support and reduced settlement. Detailed guidelines for the design of the CPRF foundation system can be found in IS: 19,117 [6].

A cushion layer is introduced in the NCPRF configuration to reduce direct interaction between the raft and pile heads. Two thicknesses are analyzed, 0.25 and 0.5 m, each with a plan dimension of 7 × 7 m. The configuration of the connecting and non-connecting CPRF system is shown in Fig. 1b. The properties used for soil material, piled-raft system, and cushion layer are listed in Tables 1 and 2, respectively. The Plastic Hardening (PH) constitutive model (a shear and volumetric hardening constitutive model) is used for the simulation of soil behavior.

Accurate mesh design is essential in finite element analysis to balance computational efficiency with solution precision [7]. In this study, a finer mesh is employed around the foundation region, particularly near the pile-soil and raft-soil interfaces, where stress gradients and dynamic interactions are most critical. The FE model (see Fig. 1a) applies fixed boundary conditions at the base to simulate rigid bedrock, preventing vertical and horizontal movements. Free-field boundaries are assigned on the sides to allow wave propagation and minimize reflections. These arrangements replicate realistic soil behavior and reduce boundary effects. For dynamic analysis, self-weight is included, and sinusoidal ground accelerations of 0.1, 0.2, 0.3 and 0.4 g are applied with an excitation frequency of 12.45 Hz, and 5% soil damping is assumed.

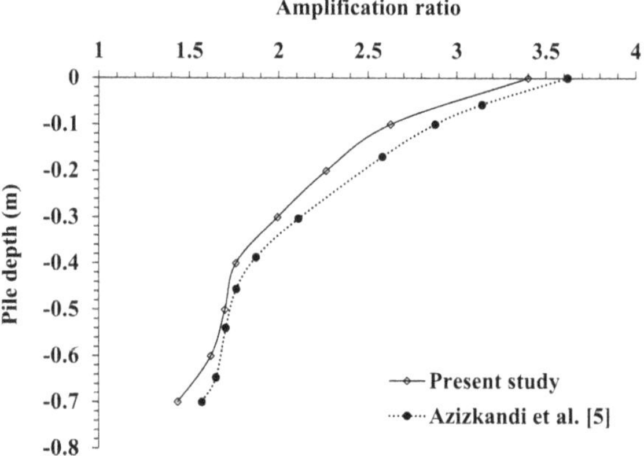

Fig. 2 Comparison of amplification profile: experimental versus numerical study

3 Results and Discussions

The FE model was validated using an experimental study consisting of centrifuge tests conducted by Azizkandi et al. [5]. Scaling factors based on centrifuge principles were used to match the prototype and model conditions. Validation focused on comparing the amplification ratio along the pile depth, shown in Fig. 2. The solid line indicates results from the present study, while the dashed line represents the centrifuge test result. Both curves follow a similar trend, with an increasing amplification ratio with depth. The close agreement confirms the accuracy of the model in simulating seismic responses.

From Fig. 3a, it can be observed that NCPRF, with a 0.5 m cushion layer, shows a decreasing amplification trend from 1.82 at 0.1 g to 1.25 at 0.4 g, highlighting effective damping. The 0.25 m cushion layer in NCPRF also reduces amplification, though less significantly. Figure 3b presents the effect of seismic acceleration on bending moments along pile elevation for both CPRF and NCPRF systems. The bending moment increases with seismic intensity, reaching approximately 35 kN/m at −2 m elevation under 0.4 g. The results confirm that cushion layers significantly reduce seismic-induced bending moments, with thicker layers offering better performance.

4 Conclusions

This study analyzed the static and dynamic behavior of CPRF and NCPRF systems using finite element modeling. CPRF systems exhibited higher bending moments and amplification factors under seismic loading. NCPRF systems, especially with thicker cushion layers, showed improved performance in reducing seismic effects. A 0.5 m cushion layer significantly lowered both bending moments and acceleration amplification. Amplification in CPRF increased with ground acceleration, while in

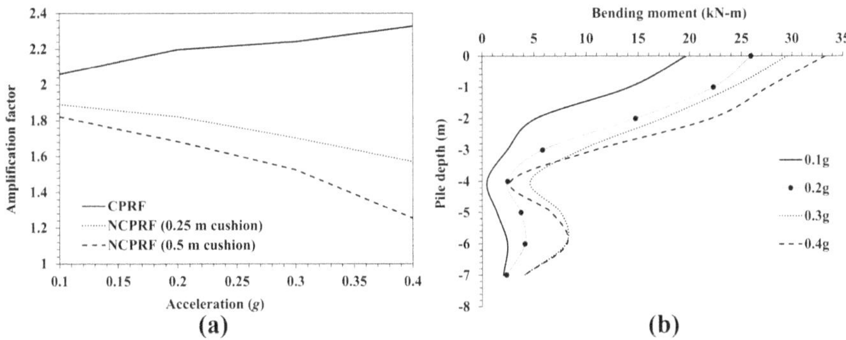

Fig. 3 Results of FE analysis: **a** amplification profile; **b** bending moment variation

NCPRF, it decreased. Overall, the NCPRF with thicker cushions is more effective for seismic mitigation.

Competing Interests The author(s) has no competing interests to declare that are relevant to the content of this manuscript.

References

1. Katzenbach R, Leppla S, Choudhury D (2016) Foundation systems for high-rise structures. CRC Press, Taylor & Francis Group, USA and UK. (ISBN: 978–1–4987–4477–5)
2. Kumar A, Choudhury D, Katzenbach R (2016) Effect of earthquake on combined pile–raft foundation. Int J Geomech 16(5):04016013
3. Bhaduri A, Choudhury D (2021) Steady-state response of flexible combined pile-raft foundation under dynamic loading. Soil Dyn Earthq Eng 145:106664
4. Patil G, Choudhury D, Mondal A (2023) Interactive iterative methodology for seismic analysis of pile-raft supported nuclear power plant structure in a conventional framework. Soil Dyn Earthq Eng 173:108109
5. Azizkandi SA, Rasouli H, Baziar MH (2019) Load sharing and carrying mechanism of piles in non-connected pile rafts using a numerical approach. Int J Civ Eng 17:793–808
6. IS 19117: Design and construction of combined piled-raft foundations. Code of Practice, Bureau of Indian Standards, New Delhi, India
7. Das T, Sharma M, Choudhury D (2023) Bearing Capacity and liquefaction assessment of shallow foundations resting on vibro-stone column densified soil in Vallur oil terminal, India. Indian Geotech J 53:1392–1413

Failure Analysis of a Large-Scale Hoarding in Mumbai: Lessons in Wind Load Design and Foundation Safety

Pranoy Debnath, Arpita Ray, and Deepankar Choudhury

Abstract This study investigates the failure of a large-scale advertisement hoarding in Mumbai, India, in 2024, which collapsed under wind loads exceeding its design capacity, along with various other reasons. A forensic analysis highlights critical inadequacies in foundation design, wind load estimation, and code compliance. Using field data and IS codal provisions, the paper identifies gaps in structural and geotechnical design practices for such hoardings in coastal urban regions. The study proposes region-specific design guidelines, emphasizing wind pressure coefficients, foundation stability, and safety margins. This work aims to improve future hoarding installations by integrating engineering design with municipal policy to ensure structural safety and public welfare.

Keywords Hoarding collapse · Wind load design · Foundation safety · Geotechnical failure analysis

1 Introduction

Hoardings, commonly known as billboards, are freestanding or wall-mounted structures used for displaying advertisements, public messages, or promotional content in urban areas. These structures are strategically positioned at high-visibility locations to maximize audience reach. Often placed in public spaces, the design and structural integrity of hoardings must adhere to both national and local regulations

P. Debnath · A. Ray · D. Choudhury (✉)
Department of Civil Engineering, Indian Institute of Technology Bombay, Powai, Mumbai, India
e-mail: dc@civil.iitb.ac.in

P. Debnath
e-mail: pranoy.debnath9@gmail.com

A. Ray
e-mail: 24d0280@iitb.ac.in

D. Choudhury
High-Rise Committee (HRC) of BMC, Mumbai, India

M. Alzaylaie et al. (eds.), *Geotechnical Innovation*, Lecture Notes in Civil Engineering 851, https://doi.org/10.1007/978-981-95-9215-9_2

to ensure public safety. The structural design should ensure stability against wind loads, seismic forces, and environmental factors. Figure 1 shows the 120 × 140 ft Ghatkopar hoarding (Mumbai, Maharashtra, India) at GRP land before its collapse in 2024.

However, design inconsistencies and a lack of proper detailed guidelines specific to regions have led to incidents of collapse of such structures and accidents as reported in the last year. One such major incident took place at GRP land, Ghatkopar East, Mumbai (in the state of Maharashtra, India) when a 120 × 140 ft. hoarding structure underwent a massive collapse on the afternoon of May 13, 2024. The 54.6 m high, newly built hoarding erected around 12 months ago for displaying advertisements along the Eastern Express Highway, came crashing down within just 8 s over an underlying petrol pump, smashing vehicles and burying people underneath. According to news reports, a large number of people either died or were injured due to this collapse. Following this incident, many such hoardings suspected of collapse were taken down. However, with increasing retail-oriented society, the decreasing number of hoardings is hardly apprehended. Also, a simple statement like the erection of a hoarding beyond the permitted size is of no substantial help. In spite of having some guidelines from the Municipal Corporation (BMC) on the allowed maximum size of 40 × 40 ft hoarding, such oversized hoarding was erected, violating BMC's norms.

Fig. 1 The 120 × 140 ft Ghatkopar hoarding at GRP land (Mumbai, India) before collapse [1, 2]

As a preliminary estimate, on inspection, it was found that necking and bending of reinforcement bars going to piles took place along with buckling of the intermediate connecting sections. The failure was mainly attributed to the overriding of the actual wind speed (24.16 m/s) way beyond the design wind speed used (13 m/s) for the foundation and in turn to the insufficient design of the hoarding structure [1]. Consequently, the foundation capacity was exceeded, and the factor of safety for overturning moments was less than 0.5, which for stable design should have been a minimum of 1.5.

Hoardings erected in coastal areas like Mumbai in India, where wind speeds are very high (design wind speed 44 m/s as given by [9]), should especially follow wind speed guidelines for design, which were primarily absent in the failed hoarding. Also, other design codes and policy guidelines inclusive but not restricted to Indian documents must be adhered to before erecting such structures some of which are [2-9], Municipal Guidelines, IRC (Indian Roads Congress) etc.

2 Site Description

The hoarding, deemed as exceeding the permissible limits, was erected behind the Police Ground Fuel station along the Eastern Express Highway in Pantnagar, Ghatkopar, Mumbai. Ghatkopar geographically lies on the western corner of the Western Ghats, marking the beginning of the Ghats from Mumbai. In the Mumbai region, particularly in Ghatkopar, the soil composition is mainly alluvial and loamy, affected by its closeness to the sea and the Deccan basalt formations that define the area. The incident of collapse occurred amidst dust storms followed by heavy rains.

3 Structure, Foundation, and Failure Mechanism

The hoarding structure used steel frames, which were designed to resist wind pressure, design wind force, and hence overturning moment on the foundation head or pile cap of 143.33 N/sq.m, 189.92 kN, and 6751.67 kN-m, respectively.

A pile foundation was selected with 5 piles each underlying the 5 columns (hence, a total of 25 piles). 600 mm diameter M35 concrete piles reinforced with rebars of 500 MPa were used, having a compressive and tensile (uplift) strength of 650 kN and 250.2 kN respectively. The pile cap depth was taken as 2.25 m. Seismic loading was not given much priority in the tabulations.

However, the above design was found highly inadequate considering the wind speed witnessed on May 13, 2024, and the IS Code guidelines. According to Bambole and Sangle [1], the overturning resistance of the foundation complying with IS 875 (Part 3): 2016 should have been 97,167 kN-m, and the induced overturning moment was 20,705.53 kN-m on the day due to a wind speed of 24.17 m/s, while it lay as low as 6748.7 kN-m in reality. As a result, the factor of safety against overturning was found to be just 0.326, whereas the minimum requirement was 1.5 (Cl. 20.1, IS 456:2000). These significant inadequacies and a clear design fallacy resulted in complete collapse, reinstated by the necking and bending of rebars and buckling of intermediate connection sections.

4 Impact of Failure

The collapse had a detrimental impact on human lives as well as property. The crash resulted in a few deaths and many injuries among people who were passing by the road during the collapse. The damage to the petrol station was also significant, with the smashing of cars in the parking lot in the vicinity. The damages in total could run into a few tens of millions of rupees. The incident highlighted the importance of the strict monitoring of compliance with guidelines for any large or small structure. Also, special attention is to be given to hoardings in areas with a larger population density and near important public utility points like petrol stations or highways. However, one of the first steps to mitigate such disasters is the development of a proper, detailed guideline to adhere to. The study presented through this paper outlines some of the codes for the design of such hoarding structures and their foundation, abiding by national and international codal provisions and policy guidelines.

5 Engineering Design Considerations

According to the latest Brihanmumbai Municipal Corporation (BMC) guidelines, the maximum permissible size for any hoarding is 40 feet in width and 40 feet in height to ensure safety and maintain urban aesthetics. After the customary dead and live load analysis (by IS 875 (Parts 1 and 2): 1987), the wind load analysis is done, which is critical since Mumbai experiences high wind speeds, especially during monsoons. In accordance with the wind load analysis code [9], the design wind speed is evaluated by multiplying the risk coefficient, terrain roughness, height factor, topography, and cyclonic importance factors by the basic wind speed (Cl 6.3). Thereafter, the design wind pressure is calculated using the normal wind pressure (0.6 times the square of wind speed), wind directionality, area averaging, and Combination factors, but has to always be more than or equal to 0.7 times the wind pressure (Cl 7.1).

Wind load can either be evaluated using Pressure coefficients (difference between external and internal pressure coefficients (C_p) is multiplied with area of the unit and design wind pressure at the height of the surface from the ground (Cl 7.3.1) to obtain total wind load) or using Force coefficients (force coefficients (C_f) multiplied by the effective frontal area (A_e) of the building or structure and by design wind pressure (p_d), gives the total wind load (F) on that specific building or structure (Cl 7.4)). While pressure coefficients apply to a specific part of a structure, they apply to a building or structure in its entirety. Total loads in each of the critical directions from all quadrants should be considered to obtain the maximum wind load. Internal air pressure in a building relies upon the degree of permeability of cladding to the flow of air, while external air pressure coefficients are defined for specific structure types. Hoardings fall under the category of 'Clad buildings' (force coefficients specified in Cl 7.4.2) for which frictional drag is calculated when d/h or d/b is more than 4 (where d, h, and b represent depth of structure (in the direction of wind), height of structure above ground level, and breadth of the structure, respectively).

Wind interference occurs when wind flow characteristics are altered due to the obstruction of an object or structure in its path, particularly in tall buildings or structures. When this altered wind subsequently impacts another structure, it generally results in an increase in wind pressures, although shielding effects may occur in closely spaced structures. A Wind Interference Factor (IF) given in IS 875 (Part 3): 2016 is multiplied by the design wind pressure to account for interference effects.

5.1 Foundation Systems

Foundation systems for hoardings are equally important while considering the stability and integrity of the whole structure as a structural component. The foundation type depends on soil conditions and hoarding height and should be designed to avoid excessive settlement or bearing capacity failure. Geotechnical characteristics of underground materials must be estimated through detailed field and/or laboratory testing as per codal provisions before foundation design. Three common types of foundations are: Shallow foundations for small to medium-sized hoardings on firm/ stiff soil, pile foundations for heavy hoarding structures having high wind loads on weak/soft soil, and anchors and micro-piles for rooftop hoardings or those on soft soil.

The calculated bearing capacity should be checked for adequacy using proper factors of safety (for sliding, overturning, etc.). Wind-induced overturning moments and lateral forces must be accounted for in foundation and pole anchoring design. The wind load and the minimum notional or crowd load produce a combined moment (M_o) and shear force (Q_o) at the base of the post of hoarding, which can be determined using the following equations.

$$M_o = \{(Wind\ Force) \times h/2\} + \{(minimum\ notional\ or\ crowd\ load) \times 1.2\} \quad (1)$$

$$Q_o = \{Wind\ Force + (minimum\ notional\ or\ crowd\ load) \quad (2)$$

where h is the height of the hoarding (or post) above the ground level. The foundation is designed with an appropriate safety factor to withstand the applied moment and shear force. The location of the rotation point of the foundation, or fulcrum, varies based on the foundation type. For a post of foundation, the fulcrum for ground resistance is assumed to be at a depth of $0.707 \times$ planting depth (P) below the

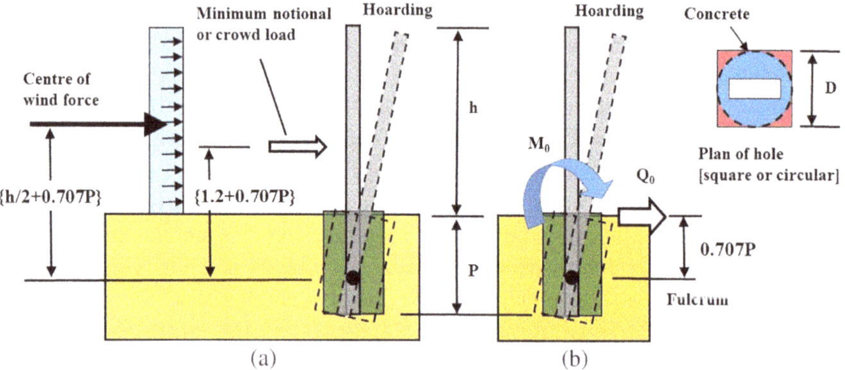

Fig. 2 Post foundation **a** Applied loads on foundations and **b** Design representations [10, 11]

ground surface (Fig. 2), and the ground resistance moment (M_g) is calculated using the following equation.

$$M_g = \left(G \times D \times P^3\right)/10 \tag{3}$$

where G represents the Ground Factor (in kN/m^2) (check Table D1 of TWf2012: 01 (revised 2014))

M_g represents the ground resistance moment (in kN-m).

P represents the planting depth (in m) of the post from the ground level.

D represents the minimum effective width (in m) of the concreted foundation.

The design must consider the overall stability safety factor, which should be at least 1.5. Hence, the stability of the post can be achieved by adjusting the post-planting depth (P) following the criteria mentioned below.

$$M_g \geq \{M_o + (0.707 \times Q_o \times P)\} \times 1.5 \tag{4}$$

Other types of foundations used for hoardings include bolted foundation (when hoardings possess a shorter design life of up to 2 years), which should have corrosion-resistant assemblies, Proprietary foundation block, which should be checked properly for sliding and overturning, and Kentledge foundation, which prevents overturning in both directions.

6 Conclusions

The design of advertisement hoardings in Mumbai like a coastal city requires a comprehensive structural, geotechnical, and environmental approach. Adherence to IS codes, wind and seismic design principles, and material selection is critical for ensuring safety. Given Mumbai's coastal location, high wind speeds, and seismic activity, hoardings must be designed to withstand extreme forces while ensuring durability with minimal environmental impact, following municipal guidelines. By integrating scientific engineering principles with municipal policies, a balance can be achieved between advertising needs and public safety.

Acknowledgement The authors gratefully acknowledge some of the information received from various agencies like Brihanmumbai Municipal Corporation (BMC), Mumbai Police, through the high-level inquiry committee investigation team appointed by Government of Maharashtra which was headed by Shri Dilip B. Bhosale (former Chief Justice of Allahabad High Court and Chairman of this Inquiry Committee), Prof. Deepankar Choudhury (Chair Professor and former Head of Civil Engg. of IIT Bombay), Ms. Ashwini Joshi (Addl. Municipal Commissioner, BMC), Shri Nikhil Gupta (Additional Director General of Police, Government of Maharashtra) and Shri Chetan Nikam (Deputy Secretary, Government of Maharashtra); and the technical investigation team's report from Veermata Jijabai Technological Institute (VJTI), Mumbai for providing valuable documents, site photo, technical inputs in support of this study. While doing this study, a portion of the technical part of the report on "High Level Inquiry Committee Headed by Justice Dilip B. Bhosale, Former Chief Justice, Allahabad High Court – Inquiring into Ghatkopar Hoarding Collapse Incident/Accident Dated 13.05.2024" has partially used which is highly acknowledged.

Competing Interests The author(s) has no competing interests to declare that are relevant to the content of this manuscript.

References

1. Bambole A, Sangle KK (2024) Investigation of "Collapse of 120 ft. × 140 ft. Hoarding at GRP Land", Ghatkopar East, Mumbai. Report prepared by Veermata Jijabai Technological Institute (VJTI), Mumbai.

2. Dilip B. Bhosale (former Chief Justice of Allahabad High Court), Deepankar Choudhury (Chair Professor and former Head of Civil Engg. of IIT Bombay), Ashwini Joshi (Addl. Municipal Commissioner, BMC), Nikhil Gupta (Additional Director General of Police, Govt. of Maharashtra) and Chetan Nikam (Deputy Secretary, GoM) (2025); "High Level Inquiry Committee Headed by Justice Dilip B. Bhosale, Former Chief Justice, Allahabad High Court – Inquiring into Ghatkopar Hoarding Collapse Incident/Accident Dated 13.05.2024", Report submitted on 7th May 2025 to the Hon'ble Chief Minister Shri Devendra Fadnavis, Government of Maharashtra, Mumbai, India, pp.1–646.

3. IS 1893-(Part 1)-2016: Criteria for earthquake resistant design of structures, part 1: General provisions and buildings. Bureau of Indian Standards, New Delhi, India.

4. IS 2062 (2011) Hot rolled medium and high tensile structural steel. Bureau of Indian Standards, New Delhi, India.

5. IS 2911-(Part 1–2)-2010: Design and construction of pile foundations—code of practice, part 1: concrete piles, section: bored cast in-situ concrete piles. Bureau of Indian Standards, New Delhi, India.

6. IS 456 (2000) Plain and reinforced concrete—code of practice. Bureau of Indian Standards, New Delhi, India.
7. IS 800 (2007) General construction in steel—code of practice. Bureau of Indian Standards, New Delhi, India.
8. IS 875-(Parts 1, 2)-1987: Code of practice for design loads (other than earthquake) for buildings and structures, part 1: Dead loads—unit weights of building material and stored materials, part 2: imposed loads. Bureau of Indian Standards, New Delhi, India.
9. IS 875-(Part 3)-2016: Code of practice for design loads (Other than earthquake) for buildings and structures—part 3: Wind loads. Bureau of Indian Standards, New Delhi, India.
10. TWf (Temporary Works forum) 2012:01, Hoardings: A guide to good practice (revised April 2014)
11. Williams BP, Waite D (1993) The design and construction of sheet-piled cofferdams. CIRIA Special Publication 95

Micropiles in Urban Infrastructure: A Case Study of Skywalk Construction Connecting Riyadh Airport Terminal with Metro Line

Haytham T. S. Hamdan and Hameeduddin Mohammed

Abstract Constructing a skywalk to connect Riyadh Airport terminal with the Metro line station over an existing parking building required an innovative and executable solution. A deep foundation micropile solution for supporting the skywalk on twin piers was engineered to overcome challenges related to constructability, geotechnical and structural design parameters, and adaptability to avoid conflicts with the functionality of the parking structure. A micropiling support system on pad foundations was designed to resist vertical, horizontal, and dynamic loads while minimizing differential settlements to support the twin piers of the skywalk. A self-drilling hollow bar system equipped with pressure grouting was used to construct 250 mm diameter micropiles. The challenge of resisting high lateral forces and displacements was addressed by adding steel pipe reinforcement to the top portion of the micropiles. Theoretical pile capacities were confirmed by performing axial and lateral pile load tests.

Keywords Micropiling · Sustainable deep foundation solutions · Pile load test · Urban infrastructure

1 Introduction

Urbanization has intensified the need for innovative foundation solutions capable of overcoming challenges such as restricted access, vibration limitations, and complex soil conditions. Among these advancements, micropiles—small-diameter drilled and grouted piles—stand out as a versatile and efficient choice for urban infrastructure projects. Their installation process is remarkably well suited to confined spaces, minimizing disturbance to surrounding environments. This makes them indispensable for

H. T. S. Hamdan (✉) · H. Mohammed
Bullivant Arabia Ltd., Riyadh, Saudi Arabia
e-mail: haythamhamdan@bullivantarabia.com

H. Mohammed
e-mail: m.hameeduddin@bullivantarabia.com

© The Author(s) 2026

M. Alzaylaie et al. (eds.), *Geotechnical Innovation*, Lecture Notes in Civil Engineering 851, https://doi.org/10.1007/978-981-95-9215-9_3

both new construction and the retrofitting or seismic strengthening of existing structures. Additionally, micropiles generate minimal noise and vibration during installation, ensuring harmony with urban settings. Combining strength, adaptability, efficiency, and resilience, micropiles provide a cutting-edge solution to the evolving demands of urban infrastructure.

When selecting a micropile foundation for a project, several critical factors must be carefully evaluated. The first consideration is the soil's load-bearing capacity, which determines the required size and depth of the piles to adequately support the imposed loads. Second, the type of rock or soil into which the piles will be installed must be considered, as this influences both the installation process and the foundation's long-term performance. The third factor is groundwater level; piles installed in areas with high groundwater must be designed to withstand hydrostatic pressure effectively. Finally, it is essential to account for the climate and weather conditions specific to the installation site. This ensures the micropile foundations are resilient to potential corrosion and thermal or moisture-related expansion and contraction, maintaining structural integrity under varying environmental conditions.

Micropiles stand out as a preferred choice in scenarios where traditional piles face limitations. They excel in various challenging conditions. Moreover, their impressive load-bearing capacities, ranging from 100 to 5000 kN, further solidify their role as a versatile and reliable foundation solution.

2 Project Description

The elevated metro stations 4G2 and 4G3 of the Riyadh Metro line are connected to the terminals of King Khaled International Airport by 200-m-long pedestrian skywalk bridges. These bridges pass over the existing car parking structures and are supported over twin piers, featuring spans ranging from 7 m up to 45 m. The bridge piers penetrate through the slabs of the parking structure to be supported at the base level, which required openings to be made in the parking structure slabs. To ensure optimal placement, the support locations were carefully chosen to avoid interference with the structural framing and traffic flow of the car park buildings.

The selection of pier foundations involved meticulous consideration of geotechnical and structural constraints. Ultimately, pad foundations supported on micropiles were identified as the most suitable solution, effectively addressing the project's unique challenges (Figs. 1, 2 and 3).

The soil beneath the car park structure consists of an alluvium layer of granular soil with a thickness ranging from 23 to 35 m, which is generally medium dense to very dense silty fine sand, becoming denser with depth. It contains interbedded layers of cemented cohesive soil. Bedrock was found starting at depths below ground level ranging from 27 to 42 m. The lithotypes described correspond to limestones, calcarenite, interbedded limestones and calcarenites, and breccia of the Sulaiy formation. No groundwater was encountered up to 60 m, the maximum drilled depth of the boreholes from ground level.

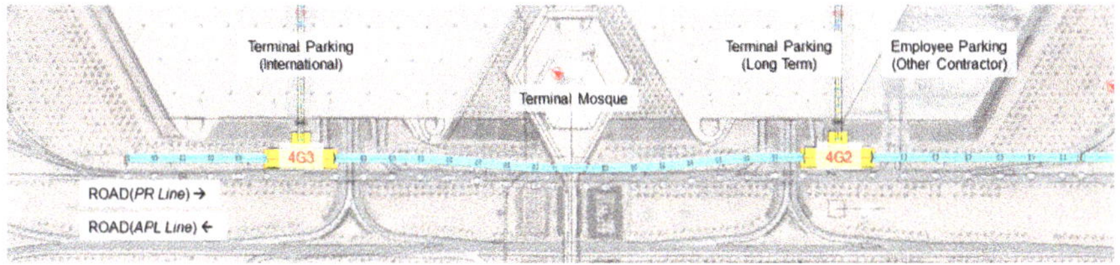

Fig. 1 Location of metro stations 4G2 and 4G3 and skywalks (connection with terminals)

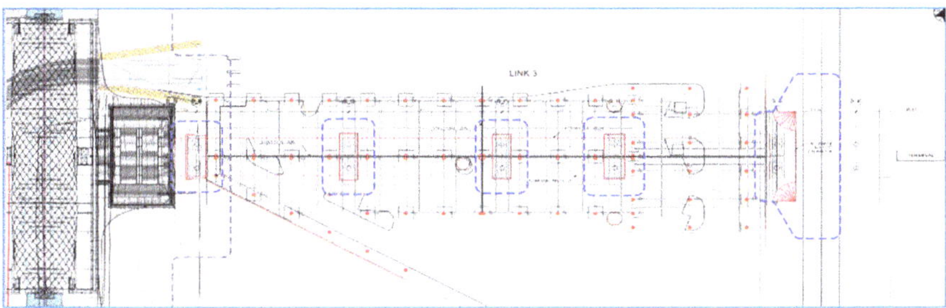

Fig. 2 4G2 Skywalk plan

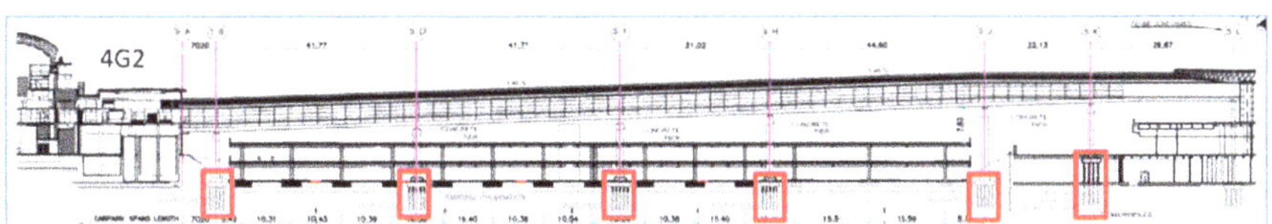

Fig. 3 4G2 Skywalk section

The number and length of micropiles were designed by considering structural loads, subsoil conditions, soil-structure interaction, group action, settlement requirements, and induced settlement on existing and adjacent structures.

Hollow bar pressure-grouted micropiles (Type-E) [2], 14 m in length and reinforced with 40/16 hollow bars having a yield strength of 583 MPa, were designed to support pier foundations. The grout body of the micropiles was 250 mm in the upper-cased portion and 225 mm in the uncased portion. Steel casing of 219.1 mm diameter and 18.8 mm wall thickness conforming to ASTM A53 Gr B was used in the upper 2.75-m portion of micropiles below respective pile caps to accommodate induced moments and shear forces due to applied lateral loads on the micropiles (Figs. 4, 5 and 6).

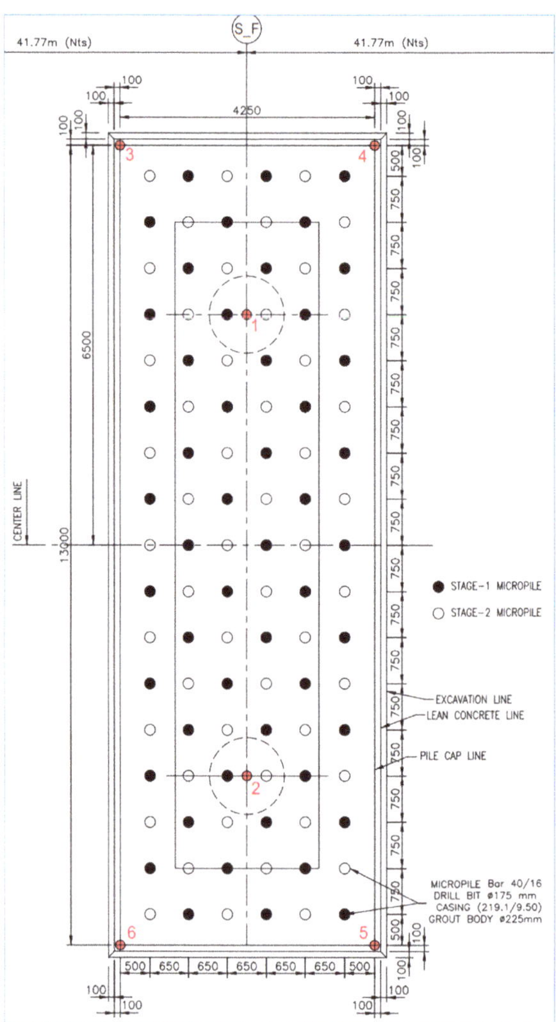

Fig. 4 Skywalk pier foundation plan

3　Design of Micropile Foundation

Micropiles were designed to support the skywalk piers' foundations. These elements transfer loads from the Skywalks into granular soil formation by means of skin friction resistance. Tip resistance was neglected for micropile design. The guidelines provided in AASHTO LRFD 2012 [2] were followed to design the micropile foundations. These micropile foundations were designed for overall stability, considering axial resistance in compression and tension, lateral resistance, group effect, and settlement. Foundation design was verified and validated for a maximum settlement of 25 mm and a maximum induced settlement on existing car park foundations of

Fig. 5 Skywalk pier section

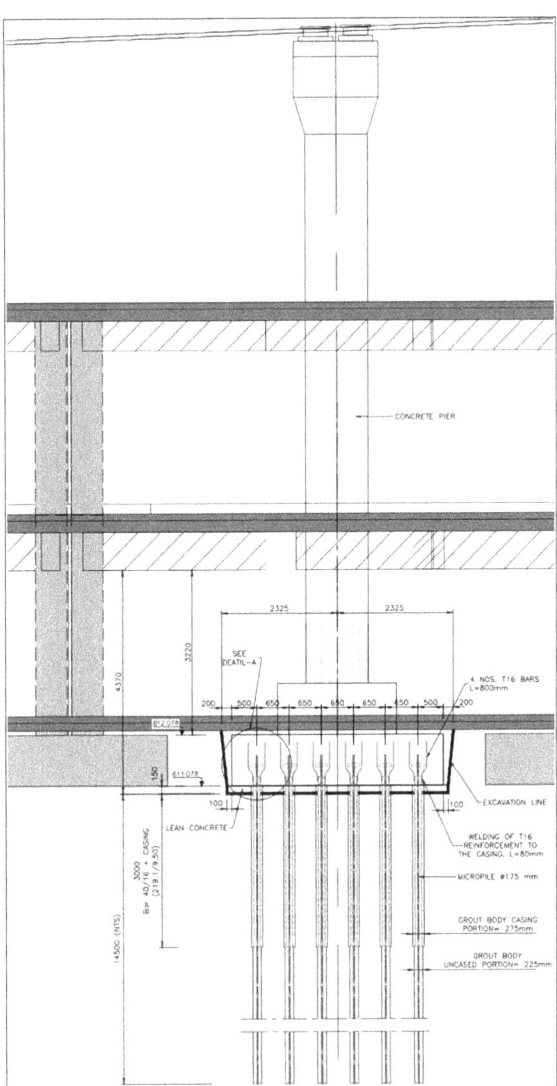

10 mm with a maximum slope of 1/500, to avoid any possible damage to the car parking structure.

It was assumed that micropiles' heads are embedded into pile caps with a degree of fixity of 100%. Horizontal loading assessment was performed following the analysis method developed by Reese (1986), which models horizontal resistance using P-y curves. Settlement assessment was performed using a resistance factor of 0.60 for axial tensile capacity based on successful static axial tensile load tests, which was used to calculate the axial tensile and compressive resistance of micropiles based on skin friction only.

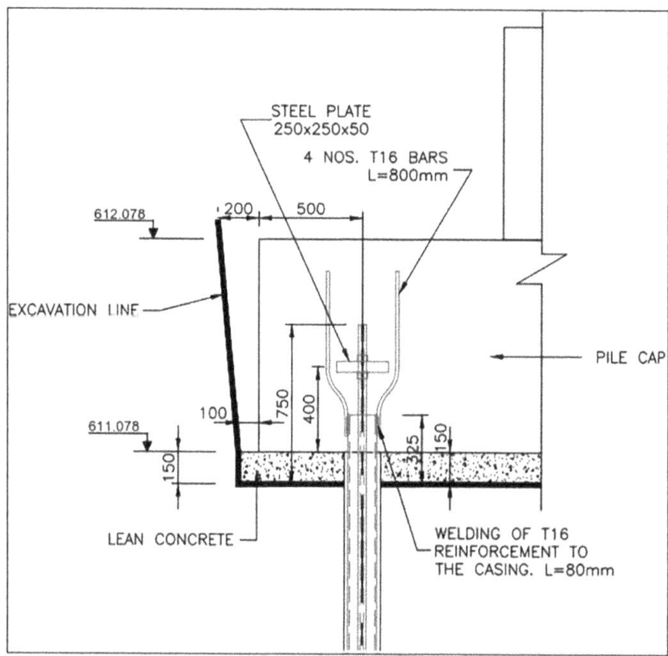

Fig. 6 Micropile head sectional detail

Structural and geotechnical axial capacities in compression and tension for micropiles were calculated as per AASHTO guidelines [2]. Nominal grout-to-ground bond strength considered in the calculations was 67 kN/m^2 as per table C10.9.3.5.2–1. Geotechnical axial capacities were analyzed by modeling a single micropile of each pier foundation in TZPILE software to determine the vertical deflections of micropiles under applied axial compressive and tensile loads. Geotechnical lateral capacities of micropiles were analyzed by modeling single micropiles in LPILE software, considering the combined effects of axial compression and lateral load, as well as axial tension and lateral load by assigning p-multipliers to micropiles based on center-to-center spacing of micropiles. The lateral deflection under each case was found to be below the allowable deflection limit of 10 mm (Tables 1 and 2).

Structural capacities of micropiles in the upper-cased portion due to combined axial and lateral loads were calculated and compared with the induced moments and shear forces in micropiles as per LPILE analyses and found to be adequate. Structural capacities of micropiles in the lower uncased portion due to combined axial and lateral loads were calculated by modeling micropiles in the spColumn software. These capacities were compared with the induced moments and shear forces in micropiles as per LPILE analyses and found to be adequate. Axial tensile and lateral load tests were carried out to verify the soil parameters used for determining skin friction and lateral response of the micropile, which were found to be satisfactory (Table 3).

Table 1 Maximum load per pile

Pier	Maximum axial compressive load	Maximum axial tensile load
	$(P_u)_{max}$	$(T_u)_{max}$
	KN	KN
S_B	257	144
S_D	298	144
S_F	330	158
S_H	323	155
S_J	352	156
S_K	308	129

Table 2 Structural capacities of micropiles

Pier	Axial compression				Axial tension			
	Max. load $(P_u)_m$	Structural capacity, R_{CU}	Result		Max. load $(T_u)_m$	Structural capacity, R_{TU}	Result	
	KN	KN			KN	KN		
S_B	257	1088	$R_{CU} > (P_u)_m$	**OK**	144	420	$R_{CU} > (P_u)_m$	**OK**
S_D	298	1088	$R_{CU} > (P_u)_m$	**OK**	144	420	$R_{CU} > (P_u)_m$	**OK**
S_F	330	1088	$R_{CU} > (P_u)_m$	**OK**	158	420	$R_{CU} > (P_u)_m$	**OK**
S_H	323	1088	$R_{CU} > (P_u)_m$	**OK**	155	420	$R_{CU} > (P_u)_m$	**OK**
S_J	352	1088	$R_{CU} > (P_u)_m$	**OK**	156	420	$R_{CU} > (P_u)_m$	**OK**
S_K	308	1088	$R_{CU} > (P_u)_m$	**OK**	129	420	$R_{CU} > (P_u)_m$	**OK**

4 Pile Load Test

To assess the load-bearing capacity and verify geotechnical parameters, an axial tension pile load test and a lateral pile load test were performed. The micropiles were tested according to the testing schedule provided by FHWA section 7.4.1 [1], and the results were checked according to the acceptance criteria provided in FHWA section 7.4.3 [1]. The axial tension test was conducted at 2.0 times the design load of 430 kN. The lateral load test was conducted for a maximum load of 48 kN. The results of these tests are shown in Figs. 7 and 8.

The maximum recorded deflection during the axial tension load test for the design load of 430 kN was measured to be less than 5.0 mm, and for the maximum load of 860 kN, it was measured to be less than 6.0 mm. The maximum recorded deflection during the lateral load test for the design load of 24 kN was measured to be less than 2.5 mm, and for the maximum load of 48 kN, it was measured to be less than 3.5 mm. Both pile load tests passed the acceptance criteria.

Table 3 Geotechnical design of micropiles for deflection

Pier	Required Pile Length below Pile Cap			Vertical Deflection in axial compression as per tzpile analyses with governing lengths						Vertical deflection in axial tension as per TZ pile analyses with governing lengths					
	Axial compression	Axial tension	Governing	Max. deflection, Y_{cm}	Allowable deflection, Y_a	Result				Max. deflection, Y_{Tm}	Allowable deflection, Y_a	Result			
	m	m	m	mm	mm					mm	mm				
S_B	13	8	**13**	1.13	10	Y_{cm}	<	Y_a	**OK**	0.52	10	Y_{Tm}	<	Y_a	**OK**
S_D	15	8	**15**	1.36	10	Y_{cm}	<	Y_a	**OK**	0.52	10	Y_{Tm}	<	Y_a	**OK**
S_F	17	8	**17**	1.57	10	Y_{cm}	<	Y_a	**OK**	0.57	10	Y_{Tm}	<	Y_a	**OK**
S_H	16	8	**16**	1.53	10	Y_{cm}	<	Y_a	**OK**	0.56	10	Y_{Tm}	<	Y_a	**OK**
S_J	18	8	**18**	1.72	10	Y_{cm}	<	Y_a	**OK**	0.56	10	Y_{Tm}	<	Y_a	**OK**
S_K	16	7	**16**	1.42	10	Y_{cm}	<	Y_a	**OK**	0.45	10	Y_{Tm}	<	Y_a	**OK**

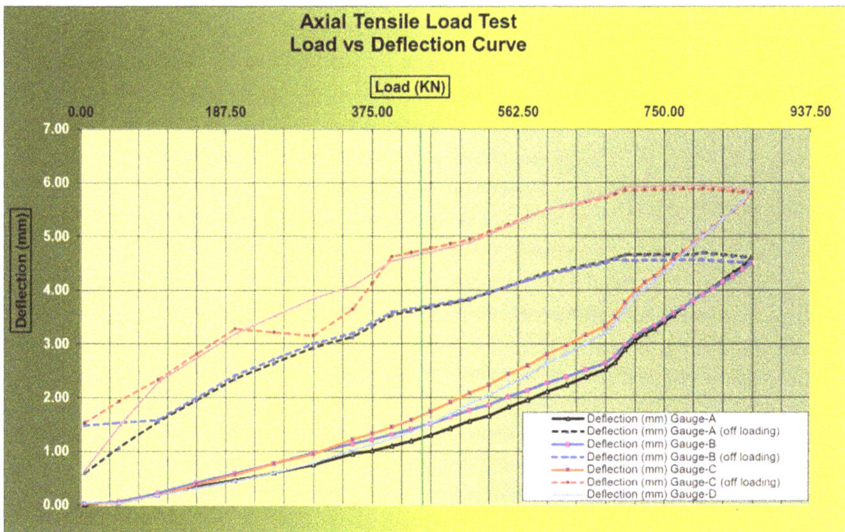

Fig. 7 Axial tension pile test results—load versus deflection curve

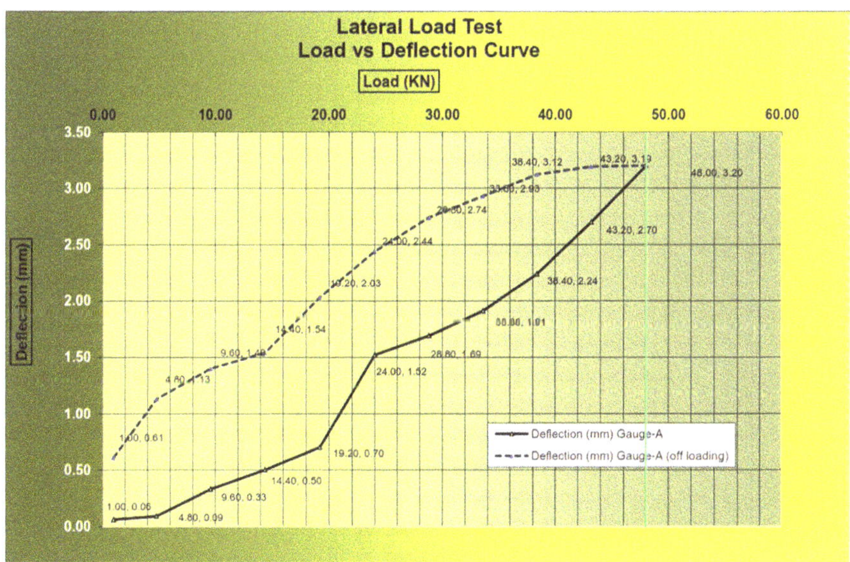

Fig. 8 Lateral load pile test results—load versus deflection curve

After installation of pier foundation micropiles, verification tension load tests on four working micropiles of the project were performed for a maximum load of 1.7 times the design load. All piles' tests passed the acceptance criteria.

5 Micropile Installation Process

Hollow bar pressure-grouted micropiles were selected as the most appropriate solution for the ground conditions encountered. A small-diameter steel bar with continuous coarse threading serves a dual purpose: it functions as both the drilling rod and grout injection conduit during installation, and later as the structural reinforcement of the pile. As drilling progresses, cementitious grout is simultaneously injected under pressure, forming a bond between the grout body and the surrounding soil. The steel bar provides tensile strength, while the hardened grout transfers axial and cyclic loads into the ground through the high bond values achieved at the grout–soil interface.

During drilling, a temporary grout with a relatively high water-to-cement ratio is circulated through the hollow bar to flush debris and stabilize the borehole. Once the target depth is reached, this flushing grout is displaced by the final mix, which is injected under controlled pressure to form the permanent pile body. The hardened grout encapsulates the central steel bar, playing a critical role in the pile's load transfer mechanism. The mix design incorporating a low water-cement ratio, microsilica, and corrosion inhibitors achieved compressive strengths above 35 MPa at 28 days in line with ASTM C109, eliminating the need for additional corrosion protection.

At the tip of the hollow bar, a disposable drill bit with pores enables grout to be injected into the surrounding soil and remains embedded once installation is complete. Drilling fluid is circulated through the hollow core, carrying cuttings out of the borehole. The type-E micropile is illustrated in Fig. 9. This construction approach significantly boosts productivity, often doubling or tripling installation speed—which in turn reduces overall project costs. Because the drilling process creates an irregular borehole surface, the grout achieves stronger mechanical interlock with the soil, thereby improving the bond along the micropile shaft and enhancing its load transfer capacity.

Following the site layout and vertical alignment of the drilling rig, installation proceeded with the hollow bar system. After grouting, a steel casing was inserted into the pile and secured in position. To complete the assembly, a square steel plate of 250 mm by 50 mm was fixed to the bar using hex nuts. Additional reinforcement was welded to the casing, forming a robust pile head that ensured effective load transfer and safeguarded against punching-type failures (Fig. 10).

Fig. 9 Type-E micropile

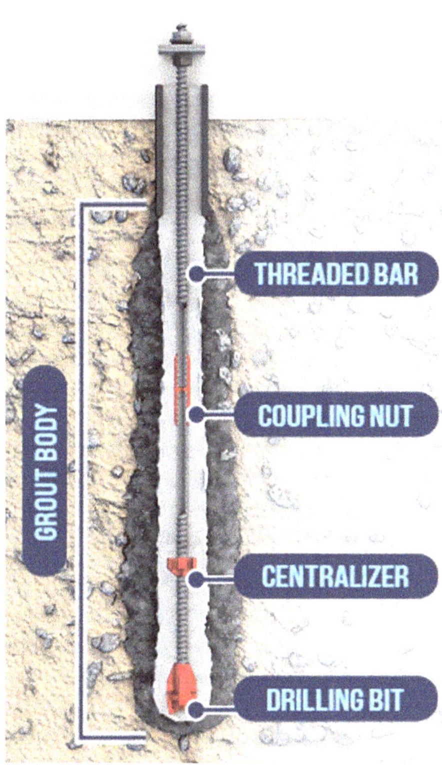

Fig. 10 Site photographs during construction of micropiles

6 Conclusions

The main conclusions that can be drawn from this Riyadh Airport Skywalk project case study are as follows:

The successful connection of Riyadh Airport terminals to metro stations 4G2 and 4G3 demonstrates the feasibility of constructing complex pedestrian infrastructure over existing facilities without compromising their functionality. The 200-m-long skywalk bridges with spans up to 45 m were successfully supported while maintaining full operation of the underlying parking structures, proving that strategic foundation placement can accommodate multi-use urban spaces.

The project overcame significant site-specific challenges including penetrating through existing parking structure slabs, working within confined spaces with restricted access, and managing construction activities in a fully operational airport environment. The solution achieved the stringent settlement requirements, ensuring no damage to the parking facility while supporting substantial skywalk loads.

Load testing validation proved critical to project success, with both axial tension tests and lateral load tests confirming the design assumptions. The measured deflections remained well within acceptable limits, providing confidence in the long-term performance of the skywalk connection. Verification tests on four working piles further validated the foundation system's reliability for this critical transportation infrastructure.

Competing Interests The author(s) has no competing interests to declare that are relevant to the content of this manuscript.

References

1. Federal Highway Administration. Micropile design and construction. Reference nanual-2005; NHI Course No. 132078; National Highway Institute, U.S. Department of Transportation
2. AASHTO LRFD bridge design specifications, Section 10.9—Micropiles

Recycling of Crushed Limestone and Tires Rubber Waste for Use as a Geotechnical Material, Initial Assessment

Saad Alsabr, Badr Almutairi, Mohamed Farid Abbas, and Sultan Almuaythir

Abstract This study investigates the recycling of crushed limestone (*CL*) from construction and demolition (*C&D*) activities and scrap rubber (*R*) from end-of-life tires to promote sustainability and circular economy principles in Saudi Arabia. The research aims to examine the physical and mechanical properties of Crushed Limestone and Rubber Mixture (*CLRM*), including specific gravity (G_s), compaction, one-dimensional compressibility, and California bearing ratio (*CBR*) testing, as a preliminary assessment for potential use as an alternative for geotechnical materials used in different applications. The aim is to determine optimal mixtures for various geotechnical applications, addressing the challenges of waste management in the context of Saudi Arabia's rapid development and Vision 2030 objectives. The design of experiment technique was employed to select the mixture doses. The findings highlight the potential of *CLRM* as a sustainable alternative to subgrades and embankments, offering a contribution to the development of sustainable construction materials, aligning with the United Nations' Sustainable Development Goals, and supporting Saudi Arabia's transition toward a more efficient and environmentally responsible economy.

S. Alsabr · B. Almutairi · M. F. Abbas (✉) · S. Almuaythir
Department of Civil Engineering, College of Engineering, Prince Sattam Bin Abdulaziz University, Al-Kharj, Saudi Arabia
e-mail: m.abbas@psau.edu.sa

S. Alsabr
e-mail: 441051474@std.psau.edu.sa

B. Almutairi
e-mail: 441051144@std.psau.edu.sa

S. Almuaythir
e-mail: s.alhomair@psau.edu.sa

M. F. Abbas
Soil Mechanics and Geotechnical Engineering Research Institute, Housing and Building National Research Center (HBRC), Giza, Egypt

© The Author(s) 2026

M. Alzaylaie et al. (eds.), *Geotechnical Innovation*, Lecture Notes in Civil Engineering 851, https://doi.org/10.1007/978-981-95-9215-9_4

Keywords Crushed limestone · Rubber waste · Permeability · One-dimensional compressibility · California bearing ratio

1 Introduction

Environmental pollution is a significant global challenge, with human activities contributing to the deterioration of air, water, and soil quality. This degradation results from various sources, including fuel consumption, solid and industrial waste, and construction and demolition (*C&D*) activities. Pollution directly impacts human health and biodiversity, threatening the balance of ecosystems [1–3].

Saudi Arabia has experienced a significant transformation driven by Vision 2030, which aims to transform the Kingdom into a global commercial and industrial center while creating an efficient and sustainable economy. One of the core objectives of Vision 2030 is the development and implementation of sustainable projects aligned with the United Nations' Sustainable Development Goals (*SDGs*), which promote the circular economy and reduce waste [4].

Various Giga projects in cities like Jeddah, Eastern Province, and particularly Riyadh, generate large amounts of waste, including crushed limestone (*CL*) from *C&D* activities. Disposing of this material in landfills contributes to visual pollution, inefficient land use, and the spread of dust storms, which pose a threat to public health [5–7].

The transportation sector is also a major contributor to waste, particularly through end-of-life tires (*ELTs*), which can no longer be reused in their original form. In countries experiencing significant population growth, such as Saudi Arabia, this issue becomes more pronounced. The current annual production of *ELTs* in Saudi Arabia exceeds 27 million tires or approximately 150,000 tons, and these values are expected to increase, posing serious concerns for the sustainability of natural resources [8, 9].

This study aims to recycle *CL* and scrap rubber (*R*) from *ELTs* to contribute to practicing sustainability and circular economy concepts. The study will investigate the main geotechnical characteristics of *CLRM*, such as specific gravity (*Gs*), compaction, hydraulic conductivity (*k*), one-dimensional compressibility, and California bearing ratio (*CBR*), to assess the potential use for different geotechnical applications. By filling the research gap on *CLRM*, this study will contribute to developing sustainable construction materials and enhancing environmental practices. Subsequent paragraphs, however, are indented.

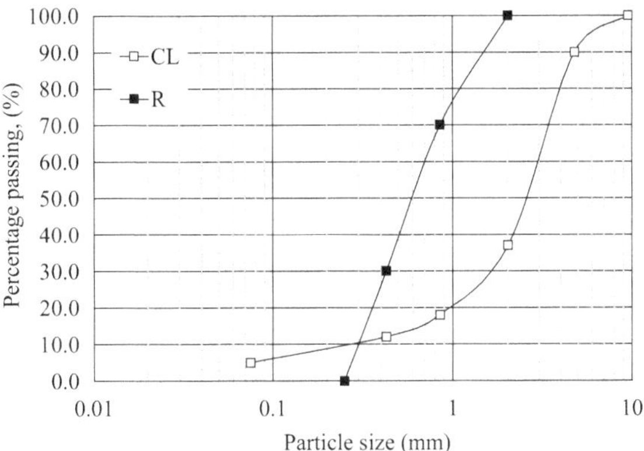

Fig. 1 Particle size distribution curves for tested materials

2 Materials Used

2.1 Crushed Limestone

The crushed limestone (CL) was obtained from a site in Riyadh and has particle sizes from 0.075 mm to 4.75 mm. It is classified as well-graded sand according to the Unified Soil Classification System. Its gradation metrics are $D_{50} = 2.65$ mm, coefficient of uniformity $= 5.64$, and coefficient of gradation $= 2.35$, as shown in Fig. 1.

2.2 Shredded Rubber Waste

Shredded rubber waste was sourced from two factories, blended in precise proportions, and processed to achieve a uniform granule size distribution ranging from 0.250 to 0.850 mm and $D_{50} = 0.60$ mm, as shown in Fig. 1.

3 Experimental Program

In this study, four test series were performed, with varying rubber content (R) as specific gravity, ASTM D854 [10], compaction, ASTM D698 [11], one-dimensional compressibility, ASTM D2435 [13], and California bearing ratio (CBR), ASTM D1883 [14]. A factorial Design of Experiments (DoE) was implemented using Minitab® to structure the testing matrix and evaluate CLRM rubber content sequence as: 0, 2.5, 5.4, 6.9, 8.3, 11.3, 14.2, 15.6, 17.1, and 20%. Throughout the study, rubber content in the $CLRM$ will be represented as 'R' followed by the percentage value; for example, '$R2.5$' indicates that the rubber content in the $CLRM$ is 2.5%.

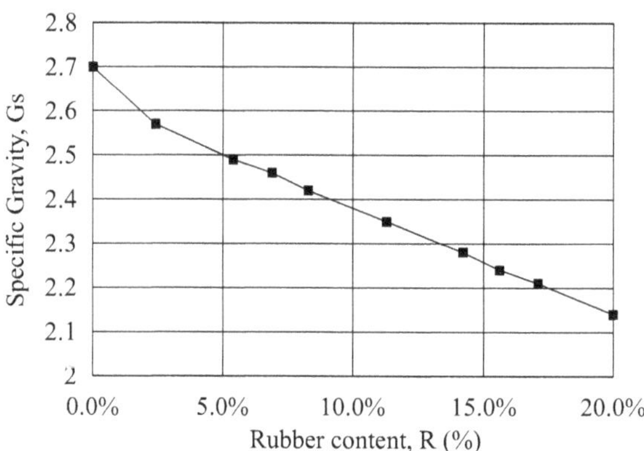

Fig. 2 Variation of *Gs* of *CLRM* with *R%*

4 Results and Discussions

4.1 Specific Gravity, **Gs**

The specific gravity of *CLRM* is reduced by increasing the *R%* as shown in Fig. 2. The monotonic reduction is attributed to an increase in *R%*, which has a low *Gs* (almost 1.15), which affects the G_s value of the overall mixture.

4.2 Compaction Characteristics

The results of the standard compaction test for the tested *CLRM* are presented in Fig. 3. This figure depicts the change in maximum dry density (*MDD*) and optimum moisture content (*OMC*) with *R%* and exhibits the effect of the *R%* on the compaction characteristics. With respect to *MDD*, adding a small dose of *R%* (i.e., 2.5%) results in a slight increase in *MDD*, followed by a monotonic reduction in *MDD* with an increase in *R%*. The initial increase of *MDD* can be credited to filling the macropores between *CL* particles by the added *R*, which has D_{50} smaller than that of *CL*. Adding more *R%* to the mixture results in replacing *CL* particles with *R* particles having a smaller value of *Gs*.

From Fig. 3, the initial increase in *R%*, i.e., up to 5%, results in a reduction in the *OMC*, followed by an increase when adding more *R%*. Based on ASTM D4609 [14], the increase in *MDD* and reduction in *OMC* are considered improvements in soil compactability, as these translate to lower costs (i.e., less compactive effort and water are required).

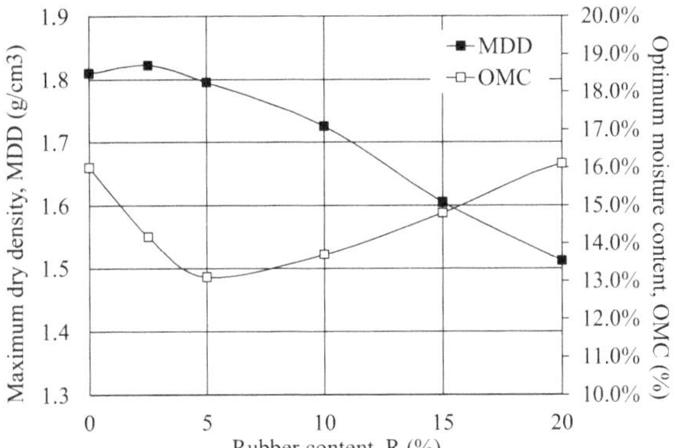

Fig. 3 Variation of MDD and OMC of CLRM with R%

4.3 One-Dimensional Compressibility Characteristics

The compressibility characteristics of soils are of great importance during construction design, as they can lead to unfavorable settlement or construction damage, which translates to cost wastage. The main compressibility parameters are compression and swell indices (*Cc* and *Cs*, respectively). So, these indices are essential for evaluating the eligibility of *CLRM* as a geotechnical material under footings. Figure 4 shows the variation of *Cc* and *Cs* values of *CLRM* with the increase in *R*%. The swell index (*Cs*) increased with the addition of *R*% up to almost 15% of rubber content, followed by a constant trend with the addition of more *R*%. Regarding the *Cc* trend of variation with adding *R*%, three trends are noted. From *R*% of 0.0% to 7.0%, almost no variation in *Cc*, followed by an increase in *Cc* with the addition of more *R*% up to 15%, where the addition of *R* results in no variation of *Cc*. Adding *R* to *CL* results in increasing the elasticity of *CLRM* up to 15.0%, where the mixture is governed by the rubber properties.

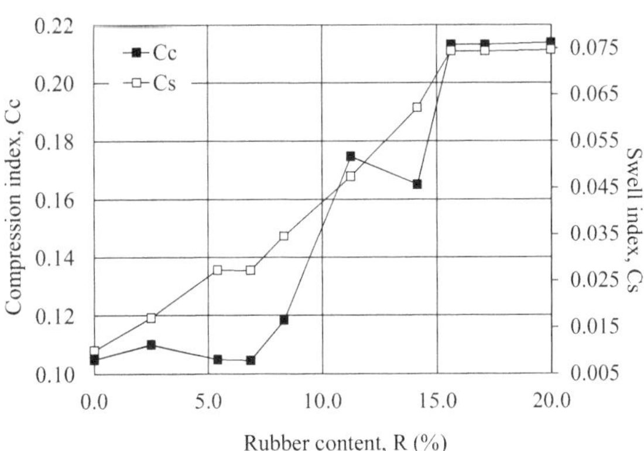

Fig. 4 Variation of *Cc* and *Cs* of *CLRM* with *R*%

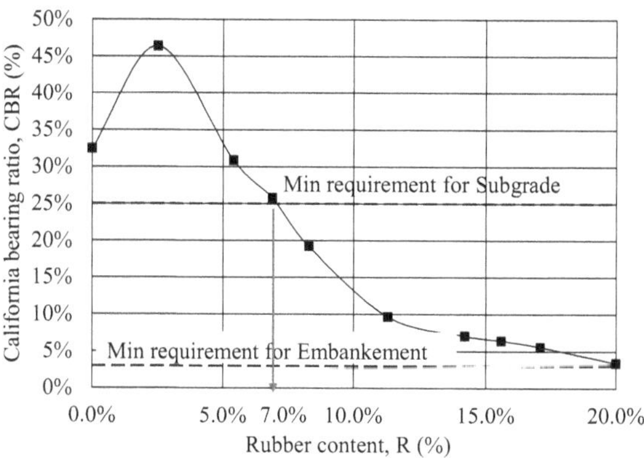

Fig. 5 Variation of *CBR* of *CLRM* with *R%*

4.4 *California Bearing Ratio,* **CBR**

The evaluation of subgrade and embankment materials is typically based on the *CBR* values. As shown in Fig. 5, the *CBR* values for the *CLRM* met the minimum allowable *CBR* local requirement for embankment material, which is >3%. In Addition, *CLRM* with *R%* ranges between 0.0% and 7.0% meet the minimum *CBR* local requirement for subgrade material, which is 25.0%.

5 Conclusion

This study investigated the viability of recycling Crushed Limestone and Rubber Mixture (*CLRM*) as a sustainable geotechnical material aligned with Saudi Arabia's Vision 2030. Through experimental evaluation of *CLRM's* physical and mechanical properties, including specific gravity, compaction, compressibility, and California Bearing Ratio (*CBR*), optimal mixtures were identified for subgrade and embankment applications.

The experimental evidence showed that adding small doses of *R%* to the mixture resulted in enhancing the compactability of the mixture, which was attributed to filling the macropores between *CL* particles. For this low range of *R%*, an enhancement in *CBR* values and maintaining mixture compressibility is noted. For a high range of *R%*, the rubber characteristics governed the mixture properties.

While this work establishes *CLRM's* laboratory-scale efficacy, an analytical study is recommended to validate its behavior under loading. Future research should also explore other properties such as hydraulic conductivity and shear strength.

References

1. Nieder R, Benbi DK, Reichl FX (2018) Soil quality and human health. In: Soil components and human health. Springer, Dordrecht. pp. 1–34. Springer Netherlands. https://doi.org/10.1007/978-94-024-1222-2_1
2. Adigun OJ, Odeleye DA (2025) The devastating consequences of environmental pollution on human health. Br J Multidiscip Adv Stud 6(1):37–46. https://doi.org/10.37745/bjmas.2022.04230
3. Gyawali K, Acharya P, Poudel D (2023) Environmental pollution and its effects on human health. Interdiscip Res Educ 8(1):84–94. https://doi.org/10.3126/ire.v8i1.56729
4. Vision 2030: Vision 2030 Annual report. Saudi Arabia, pp. 4–6. https://www.vision2030.gov.sa
5. Kibriya T (2017) Crushed limestone waste as supplementary cementing material for high strength concrete. Am J Civ Eng Arch 5(3):93–97. https://doi.org/10.12691/ajcea-5-3-3
6. Mutaz E, Dafalla M, Al-Mahbashi AM, Serati M (2024) Utilizing crushed limestone as a sustainable alternative in shotcrete applications. Materials 17(7):1–15, Article 1486. https://doi.org/10.3390/ma17071486
7. Alnuaim A, Al-Mahbashi AM, Dafalla M (2022) Utilizing Tunnel Boring Machine (TBM)-crushed limestone as a construction material. Materials 15(21):1–18, Article 7569. https://doi.org/10.3390/ma15217569
8. Tasalloti A, Chiaro G, Murali A, Banasiak L, Palermo A, Granello G (2021) Recycling of End-of-Life Tires (ELTs) for sustainable geotechnical applications: A New Zealand perspective. Applied Sci 11(17):1–15, Article 7824. https://doi.org/10.3390/app11177824
9. Balabel A, Almujibah H (2022) Towards sustainable transportation: The development of hyperloop technology in Saudi Arabia. World J Eng Technol Res 2(1):1–11. https://doi.org/10.53346/wjetr.2022.2.1.0032
10. ASTM D854: Standard test methods for specific gravity of soil solids by water pycnometer
11. ASTM D698: Standard test methods for laboratory compaction characteristics of soil using standard effort (12,400 ft-lbf/ft3 (600 kN-m/m^3))
12. ASTM D2435: Standard test methods for one-dimensional consolidation properties of soils using incremental loading
13. ASTM D1883: Standard test method for CBR (California Bearing Ratio) of laboratory-compacted
14. ASTM D 4609: Standard guide for evaluating effectiveness of chemicals for soil stabilization

Bored Pile Capacity Prediction Based on CPT Data Using Machine Learning Models

Abdulhakim Mawas and **Omar Hamza**

Abstract This study investigates the prediction of bored pile load–displacement behaviour using machine learning (ML) models. The models were trained on a dataset derived from ground investigations and 28 full-scale uplift pile tests and validated against nine independent field tests. Key input features included pile geometries, cone tip resistance (qc), sleeve friction (fs) and pile head displacement (δ). Where direct CPT data was unavailable, parameters were derived from standard penetration test (SPT) data using established correlations. Five models were developed and compared: random forest (RF), gradient boosting (GB), polynomial regression (PR), multilayer perceptron (MLP) and linear regression (LR). Model performance was evaluated using the coefficient of determination (R^2), mean squared error (MSE), root mean squared error (RMSE) and mean absolute error (MAE). Ensemble models (RF and GB) achieved the highest predictive accuracy ($R^2 \approx 0.999$), with validation results supporting their generalisation to unseen field data. Feature importance analysis confirmed that pile geometry and CPT parameters significantly influence uplift capacity. The findings demonstrate that machine learning models may supplement traditional bored pile design practices, especially when traditional data is limited.

Keywords Bored piles · Uplift capacity · Skin friction · Machine learning · Random forest · Gradient boosting · Multilayer perceptron · In-situ testing · Cone penetration test (CPT)

Glossary of Parameters and Abbreviations

A	Pile Area
ANN	Artificial Neural Network
CPT	Cone Penetration Test
CPTu	Piezocone Penetration Test
D	Pile diameter

A. Mawas (✉) · O. Hamza
College of Science and Engineering, University of Derby, Derby, UK
e-mail: 100609508@unimail.derby.ac.uk

© The Author(s) 2026

M. Alzaylaie et al. (eds.), *Geotechnical Innovation*, Lecture Notes in Civil Engineering 851, https://doi.org/10.1007/978-981-95-9215-9_5

δ (**delta**)	Pile head displacement
$\mathbf{f_s}$	Sleeve friction
$\mathbf{f_{s,av}}$	Average sleeve friction along the embedded length
$\mathbf{f_{s\_i}}$	Sleeve friction for layer i
GB	Gradient Boosting
GP	Genetic Programming
$\mathbf{k_s}$	Skin friction factor
L	Pile embedded length
LR	Linear Regression
MAE	Mean Absolute Error
ML	Machine Learning
MLP	Multilayer Perceptron
MSE	Mean Squared Error
N-values/N60	Standard Penetration Test blow counts
P	Applied uplift load
PP	Pile Perimeter
PR	Polynomial Regression
Pu	Ultimate pile uplift capacity
$\mathbf{q_c}$	Cone tip resistance
$\mathbf{q_{c,av}}$	Average cone tip resistance along embedded length
$\mathbf{q_{c\_i}}$	Cone tip resistance for layer i
$\mathbf{q_e}$	Equivalent or average layers cone resistance
$\mathbf{R^2}$	Coefficient of Determination
RF	Random Forest
RMSE	Root Mean Squared Error
RNN	Recurrent Neural Networks
SPT	Standard Penetration Test
SVM	Support Vector Machines
u	Pore water pressure
$\mathbf{z_i}$	Thickness of layer i

1 Introduction

Accurate prediction of the capacity of bored piles is paramount in geotechnical engineering, particularly for structures like transmission towers subjected to significant uplift forces. Consequently, a thorough understanding of the load–displacement relationship is indispensable for a comprehensive performance analysis of the foundation system and the superstructure it supports. Predicting the capacity and load–displacement behaviours of bored piles has historically relied on a combination of analytical, empirical and experimental approaches. Each of these methodologies offers unique advantages and limitations in addressing the complexities of soil-pile interaction.

Empirical approaches for predicting bored pile load–displacement behaviour often rely on correlations established between the results of in-situ soil tests, such as the standard penetration test (SPT) and the cone penetration test (CPT), and the observed performance of piles. The CPT is a widely adopted in-situ testing technique, valued alongside the standard penetration test (SPT) for its efficiency and ability to provide continuous subsurface data, including tip resistance (qc), sleeve friction (fs) and optionally pore pressure (u). Since its origins as the 'Dutch Cone' in the 1930s–40 s and subsequent evolution to modern electronic piezocones (CPTu), CPT data has been extensively used for soil stratification and estimating geotechnical parameters [1–4].

Various CPT-based methods have been developed for predicting the capacity of piles. These methods, such as those proposed by Schmertmann [5], Bustamante and Gianeselli [6], and Eslami and Fellenious [7], utilise the continuous measurements of cone tip resistance (qc) and sleeve friction (fs) obtained during CPT soundings to estimate the ultimate bearing capacity of piles. However, CPT-based methods for estimating pile capacity have shown considerable variability, with predictions sometimes differing by as much as 300% [5, 8]. Furthermore, many traditional methods are based on laboratory models, and there is a recognised need for more validation using full-scale tests, especially for uplift conditions, to better understand actual pile behaviour.

In recent decades, machine learning (ML) and artificial intelligence (AI) techniques have gained prominence in geotechnical engineering. These data-driven approaches offer powerful capabilities to model the complex, non-linear relationships between in-situ test data and soil or pile behaviour, potentially overcoming the limitations of traditional empirical methods. Early work demonstrated the potential of using ANNs to interpret CPT data for soil property prediction [9] leading to increased research into applying ML/AI.

The ability of ML/AI models to discern complex patterns in data has led to their increasing application in pile foundation analysis using CPT results. In addition to ANNs, researchers have employed various techniques, including support vector machines (SVM), genetic programming (GP), and ensemble methods like random forest (RF). Several researchers [10, 11] used ANNs with CPT data to investigate load-settlement behaviour. Other studies [12–14] applied ANNs, recurrent neural networks (RNN) and neuro-fuzzy models to predict pullout capacity based on lab-scale tests and CPT data. A combined approach was presented in [15]—combined polynomial neural networks with genetic algorithms for pile capacity prediction using CPT/CPTu data. Others [16] have used SVM or GP combined with ANNs [17] for similar purposes. The utility of ANN backpropagation was demonstrated in [18] for predicting SPT N60 values from CPT data. These studies collectively illustrate the potential of ML to enhance predictive capabilities compared to traditional methods.

This study contributes to this area by investigating the effectiveness of various ML models, specifically random forest (RF), gradient boosting (GB), polynomial regression (PR), multilayer perceptron (MLP) and linear regression (LR), for predicting the load–displacement behaviour of bored piles under uplift loading. A key characteristic of this research is the utilisation of CPT parameters (qc and fs—derived from in-situ testing) as well as 28 existing full-scale uplift test records. Due to the limited number of field tests, data augmentation techniques were applied to the derived dataset to enhance model training. The primary objective is to evaluate and compare the predictive performance of these ML models using this derived and augmented dataset, validated against nine independent full-scale test results, thereby assessing the viability of this approach for practical engineering applications.

Unlike earlier research focusing mostly on compression or laboratory-scale pullout behaviour, this work is supported by full-scale uplift tests. It provides a structured comparison of multiple ML models under real-world data constraints. The results offer insights into how machine learning can be realistically applied to pile design and assessment when traditional data is scarce.

2 Methodology

This section outlines the procedures followed in this study, encompassing data acquisition, preparation, augmentation, model development and evaluation for predicting the uplift load–displacement behaviours of bored piles—as shown in Fig. 1.

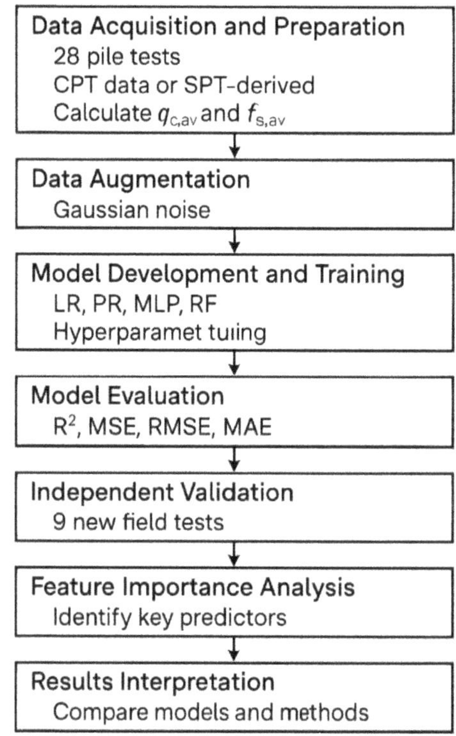

Fig. 1 Methodology adopted for the study

2.1 Data Acquisition and Preparation

Following a project-specific ground investigation that revealed silty sand, 28 full-scale uplift tests were conducted on bored piles of different lengths and diameters. The ground investigation data and the results from these pile tests constituted the fundamental dataset for this study. A critical aspect of this work is the derivation of the primary input features. The pile head displacement (δ) was treated as an input feature. The models predict the ultimate pile capacity (Pu) based on pile geometry, CPT parameters and displacement under applied uplift load.

Direct cone penetration test (CPT) measurements were not fully available for all test cases. Where missing, CPT parameters, specifically tip resistance (qc) and sleeve friction (fs), were *derived* from available standard penetration test (SPT) N-values recorded during the original site investigations.

This derivation utilised well-established SPT-CPT correlations appropriate for the predominant soil type identified at the test sites (primarily silty sand), specifically employing the relationship proposed in [4].

Following the approach suggested by [11], the derived qc and fs values were averaged along the embedded length of each pile using the following equations (based on the principle of weighted average).

$$q_{c,av} = \frac{\sum q_{c\_i} \times z_i}{\sum z_i} \tag{1}$$

$$f_{s,av} = \frac{\sum f_{s\_i} \times z_i}{\sum z_i} \tag{2}$$

where q_{c_i} and f_{s_i} are the derived cone tip resistance and sleeve friction for layer i, respectively, and z_i is the thickness of that layer.

The final input feature set for the machine learning models consisted of these averaged derived CPT parameters ($q_{c,av}$, $f_{s,av}$) along with the pile geometry (embedded length L and diameter D). The averaged value ($q_{c,av}$) represents the equivalent cone resistance (q_e) used in all model inputs and equations. The target variable for prediction was the measured pile head displacement (δ) corresponding to various applied uplift loads.

2.2 Data Augmentation

The initial dataset derived from the 28 full-scale tests was relatively small for training robust machine learning models. To address this limitation and enhance the dataset size, data augmentation techniques were employed. Specifically, Gaussian noise was injected into the original features (qc, fs, L, D, Load) before splitting into training and testing sets. Outliers were removed using Z-score thresholding. The final augmented

Table 1 Statistical comparison of origin and augmented data

		qc (MN/m^2)	fs (MN/m^2)	L (m)	PP (m)	P (kN)	δ (mm)
Mean	Origin	13.423	0.268	14.569	3.666	2095.350	4.807
	Augmented	14.309	2.862	15.082	3.824	2891.925	3.906
Standard error	Origin	0.273	0.005	0.216	0.065	108.558	0.519
	Augmented	0.196	0.039	0.164	0.043	91.381	0.175
Standard deviation	Origin	2.990	0.060	2.362	0.712	1189.192	5.683
	Augmented	4.379	0.876	3.677	0.968	2043.338	3.904
Sample variance	Origin	8.943	0.004	5.577	0.507	1,414,1⁄6.50	32.297
	Augmented	4.379	0.876	3.677	0.968	2043.338	3.904
Kurtosis	Origin	0.122	0.122	0.323	0.036	5.456	14.993
	Augmented	–0.288	–0.288	–0.780	–1.199	0.235	1.954
Skewness	Origin	–0.372	–0.372	0.372	1.045	2.345	3.237
	Augmented	0.637	–0.637	0.436	0.328	1.180	1.686

dataset of 500 samples preserved similar statistical properties to the original 28 samples, minimising risks of overfitting or data leakage. After cleaning, the data was divided into 80% for training and 20% for testing.

This involved adding small, randomly generated values drawn from a Gaussian (normal) distribution with a mean of zero and a small standard deviation to the original data points. This process generates synthetic, yet statistically similar, data points, effectively increasing the size and variability of the training dataset without requiring additional field tests. The effectiveness of this augmentation was assessed by comparing the statistical properties (mean, variance, skewness, kurtosis) of the augmented dataset with the original dataset (as summarised in Table 1).

Table 1 presents descriptive statistics (count, mean, standard deviation, min, max, quartiles) for the input features (P: load, δ: displacement, PP: pile perimeter, L: length, fs: sleeve friction, qc: tip resistance) based on the original 28 data points. This provides an overview of the range and distribution of the raw data used for derivation and augmentation.

2.3 Data Pre-processing

Following data augmentation, a rigorous pre-processing phase was undertaken to ensure the quality, consistency and suitability of the dataset for the machine learning models. This involved several key steps:

Data Diagnosis: The dataset was analysed to identify and handle potential issues such as noise, outliers and missing data. Statistical properties were examined, and visualisation techniques like histograms, boxplots and scatter plots were used to understand data distribution and relationships. Figure 2 displays histograms for each feature in the dataset. These histograms visually represented the distribution of values for load, displacement, perimeter, length, derived fs and derived qc, helping to understand data skewness or normality after augmentation.

Data Scaling: As a final preparation step, the features in the dataset were scaled. This is crucial because machine learning algorithms can be sensitive to features with vastly different numerical ranges. Both normalisation (scaling data to a fixed range, typically 0–1) and standardisation (scaling data to have zero mean and unit variance) were considered to transform the data, ensuring that all features contributed appropriately during model training. The choice between normalisation and standardisation often depends on the specific algorithm and data distribution.

2.4   Machine Learning Models

Five different machine learning algorithms were selected to model the pile load–displacement relationship and compare their predictive performance:

Linear Regression (LR): A fundamental regression model establishing a linear relationship between the input features (derived CPT parameters) and the target variable (pile displacement/capacity).

Polynomial Regression (PR): An extension of linear regression that models non-linear relationships by fitting the data to a polynomial equation of a determined degree. Model selection criteria like BIC/AIC were considered when choosing the optimal polynomial degree.

Multilayer Perceptron (MLP): A type of feedforward artificial neural network (ANN) consisting of interconnected nodes (neurons) organised in layers (input, hidden, output) It is capable of learning complex, non-linear patterns. The MLP architecture used in this study comprised two hidden layers with 50 neurons each, employing ReLU activation. As it involves multiple hidden layers, this network qualifies as a deep feedforward neural network.

Random Forest (RF): An ensemble learning method that constructs multiple decision trees during training and outputs the average prediction (for regression) of the individual trees. It is known for its robustness and ability to handle high-dimensional data.

Gradient Boosting (GB): Another powerful ensemble technique that builds models sequentially, with each new model attempting to correct the errors made by the previous ones. Specific parameters configured included the loss function, learning rate, number of estimators (trees), subsample fraction, split criterion, minimum samples for split/leaf, maximum tree depth and parameters for early stopping.

These models were implemented using standard Python libraries, including Scikit-learn for LR, PR, RF and GB, and Keras/TensorFlow for the MLP model.

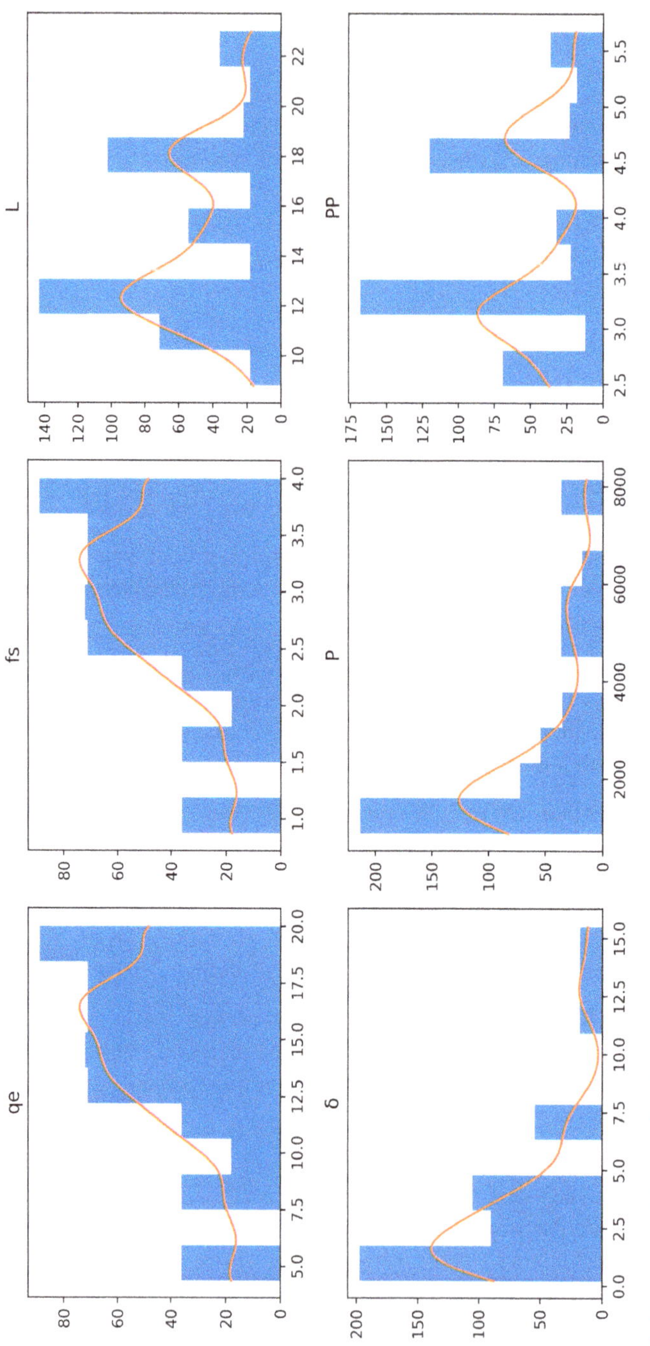

Fig. 2 Histogram of feature distributions

2.5 Model Training and Hyperparameter Tuning

The augmented dataset was appropriately partitioned into training and testing sets, 80% and 20% consequently, to train the models and evaluate their generalisation performance on unseen data from the augmented set. Standard data scaling techniques (e.g. standardisation or normalisation) were applied to the features before model training, particularly for MLP and PR/LR.

Hyperparameter tuning was performed to optimise the performance of models like MLP, RF and GB. While specific methods (e.g. grid search, randomised search) are not detailed here, tuning typically involves adjusting parameters such as the number of neurons/layers (MLP), the number of trees and tree depth (RF, GB), learning rates (GB, MLP), etc., to achieve the best performance on a validation subset of the training data or through cross-validation.

2.6 Model Evaluation and Validation

The predictive performance of the trained models was assessed using standard regression metrics:

Coefficient of Determination (R^2): Indicates the proportion of the variance in the target variable predictable from the input features. Values closer to 1 indicate a better fit.

Mean Squared Error (MSE): The average of the squares of the errors between predicted and actual values. Lower values are better.

Root Mean Squared Error (RMSE): The square root of MSE, providing an error metric in the same units as the target variable. Lower values are better.

Mean Absolute Error (MAE): The average of the absolute differences between predicted and actual values. Lower values are better.

Crucially, final model validation was conducted using an independent dataset comprising results from 9 full-scale uplift tests that were *not* included in the original 28 tests used for derivation and augmentation. This step provided a critical assessment of how well the models generalise to entirely new, unseen field data.

2.7 Feature Importance Analysis

To understand which input parameters most significantly influenced the predictions of the best-performing models (identified as RF and GB), a feature importance analysis was conducted. This typically involves using built-in methods within the ensemble models (like feature permutation importance or Gini importance for tree-based models) to quantify the relative contribution of each input feature ($q_{c.av}$, $f_{s.av}$, L, D, Load) to the prediction of pile head displacement.

3 Results

3.1 Model Performance Evaluation

The predictive performance of the five machine learning models—linear regression (LR), polynomial regression (PR), multilayer perceptron (MLP), random forest (RF) and gradient boosting (GB)—was assessed using standard statistical metrics. A comprehensive comparison of the models based on R2, mean squared error (MSE), root mean squared error (RMSE) and mean absolute error (MAE) is detailed in Table 2. Based on these metrics, the ensemble models, RF and GB, achieved outstanding performance, with R2 values approaching 1.0 (reported as 1 in the table) on both the training and testing sets, accompanied by low MSE values. This indicates an excellent fit for the data and strong generalisation capabilities within the augmented dataset. The Polynomial Regression model also performed well, achieving a test R2 of 0.87. The MLP model yielded a test R2 of 0.82, while the Linear Regression model achieved a test R2 of 0.79, demonstrating acceptable but lower predictive power compared to the ensemble and polynomial models.

3.2 Feature Importance

Analysis was conducted to determine the relative influence of the input features (derived $q_{c.av}$, derived $f_{s.av}$), pile length L, pile perimeter PP and pile head displacement (δ = delta) on the prediction of skin friction capacity. Figure 3 presents a feature correlation heatmap and feature importance results. The results indicate that cone sleeve friction (fs) and cone tip resistance (qc) are the most significant features influencing the predictions made by the models. While fs fundamentally governs unit skin

Table 2 Statistical model results for the best model selection

Model	Train MSE	Test MSE	Train R^2	Test R^2
Gradient boosting	864.147	1681.13	1	1
Random forest	3350.09	817.49	1	1
Polynomial regression	506,877.5	577,785.68	0.88	0.87
MLP regressor	624,311.91	808,433.18	0.85	0.82
Linear regression	812,144.48	942,376.50	0.80	0.79

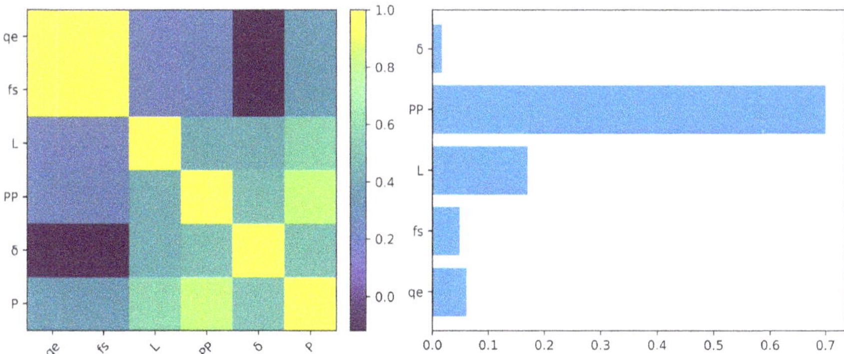

Fig. 3 Correlation matrix (left) and importance feature (right)

friction, the feature importance analysis highlighted pile perimeter (PP) and length (L) as the dominant predictors in this dataset. This is due to the significant variation in pile diameters included in the study, which leads to large changes in total shaft area and, consequently, overall capacity, overshadowing the relative variations in the derived fs values.

3.3 Predicted Versus Actual Values

Graphical comparisons between the values predicted by the models and the actual values observed in the dataset were generated. Figure 4 presents scatter plots comparing the predicted pile capacity with the actual measured capacity for each model using the test data. These plots visually confirmed the strong correlation (points clustering tightly around the identity line) for the RF and GB models, reflecting their high R2 values. The plots for PR, MLP and LR showed progressively more scatter, consistent with their lower R2 scores. A comparison chart of R2 values further highlighted the superior performance of RF and GB.

3.4 Derived Predictive Equations

Based on the regression analyses, specific equations for predicting the ultimate pile uplift capacity (Pu) were derived from the polynomial regression (PR) model and linear regression (LR) model.

The polynomial regression model for predicting Pu is given as Eq. (3).

$$Pu = 11220.1604 - 115.2446 \cdot qe - 23.0489 \cdot fs - 632.1962 \cdot L$$
$$- 4295.0397 \cdot PP + 453.2061 \cdot \delta - 1\,6820 \cdot qe2 - 0.3364 \cdot qe \cdot fs$$

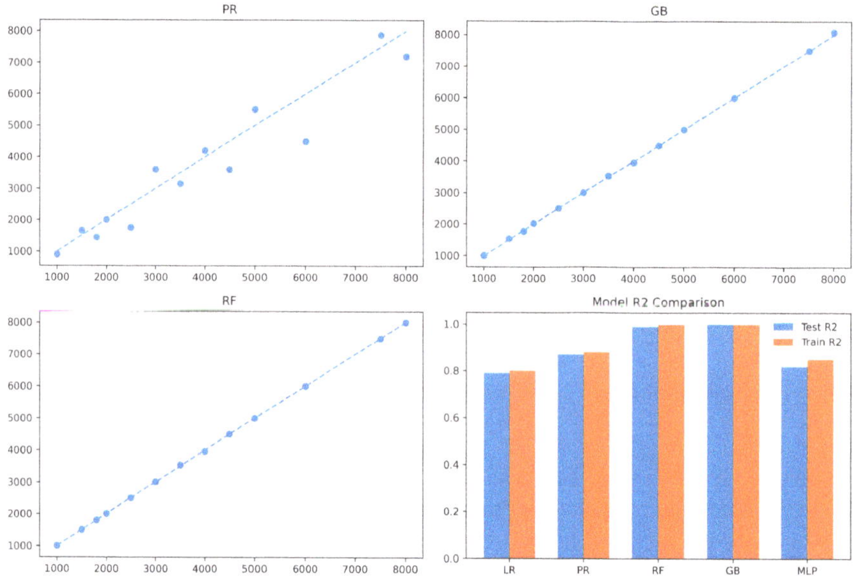

Fig. 4 Scatter graphs showing predictive vs. actual data for models. The bottom right graph shows Models R2 Comparison for the test and train datasets

$$- 0.2283 \cdot qe \cdot L + 65.4704 \cdot qe \cdot PP + 16.7994 \cdot qe \cdot \delta - 0.0673 \cdot fs$$
$$- 0.0457 \cdot fs \cdot L + 13.0941 \cdot fs \cdot PP + 3.3599 \cdot fs \cdot \delta + 19.4209 \cdot L$$
$$+ 60.5972 \cdot L \cdot PP - 38.7188 \cdot L \cdot \delta + 504.9794 \cdot PP$$
$$- 42.0240 \cdot PP \cdot \delta + 12.9344 \cdot \delta \tag{3}$$

where: PP is pile perimeter, L is embedded pile length, fs is sleeve friction unit, qc is the tip cone resistance unit, and δ is the pile head displacement.

This model achieves an R^2 of 0.88 for the training set and 0.87 for the testing set, indicating a strong fit. The inclusion of polynomial terms allows the model to capture non-linear relationships between the independent variables and the dependent variable (Pu). Furthermore, the similarity in R^2 values between the training and testing datasets suggests minimal overfitting, implying that the model generalises well to unseen data.

However, the equation lacks physical intuition due to its complexity and numerous terms. In contrast, a simpler linear regression model provides a more interpretable alternative with approximately 80% accuracy (Eq. 4).

$$Pu = -5460.6 + 81.89 \cdot qe + 16.37 \cdot fs + 116.33 \cdot L$$
$$+ 1333.64 \cdot PP + 67.87\delta \tag{4}$$

where qe is the equivalent or average layers cone resistance.

3.5 *Independent Validation Against Field Tests*

To assess the models' ability to generalise to entirely new field data, their predictions were compared against the measured results from 9 independent full-scale uplift tests not used during training or augmentation. The results indicated good agreement between the model predictions, particularly for RF, and the actual site measurements. Tables 3 and 4 provide a detailed comparison between the predicted capacities/ displacements and the measured field results, offering a validation of the models against real-world data. For the RF model, the prediction accuracy (ratio of predicted to actual capacity or displacement, though the specific metric is not defined) ranged from 101 to 102% for piles with a 1 m diameter and from 83 to 102% for piles with a 1.5 m diameter, suggesting generally high accuracy in real-world scenarios.

Table 3 Ultimate uplift pile capacity prediction for the four ML models compared with the field test results of nine piles

Pile id	L (m)	PP (m)	δ (mm)	Ultimate uplift load capacity, Pu (kN)				
				Field	RF	PR	GB	LR
1	16.7	4.71	1.10	3029	2729	3583	3514	3787
2	12.9	3.14	0.40	956	961	777	1749	861
3	12.2	4.71	1.02	3029	3087	4751	2837	4333
4	09.0	3.14	0.06	1433	1458	1264	1969	997
5	12.0	4.71	0.74	3711	3091	3943	2795	3831
6	09.7	2.51	0.79	1022	1889	1316	1274	533
7	12.0	4.71	1.31	3711	3087	3824	2916	3762
8	12.5	4.71	0.43	3711	3309	3920	4889	3846
9	11.9	3.14	1.02	1523	1556	1332	1465	1518

Table 4 The ratio (%) of ML predicted to actual ultimate uplift capacity for the nine piles used for validation

Pile id	Pile diameter (m)	RF/Actual	PR/Actual	GB/Actual	LR/Actual
1	1.5	90	118	116	125
2	1	101	81	183	90
3	1.5	102	157	94	143
4	1	102	88	137	70
5	1.5	83	106	75	103
6	0.8	185	129	125	52
7	1.5	83	103	79	101
8	1.5	89	106	132	104
9	1	102	88	96	100

3.6 Comparison with Existing Estimations Methods

The results obtained from field tests were compared with pile capacity estimations derived from methods proposed by other researchers, as presented in Table 5 and Fig. 5. The results indicate that the current study's predictions are more conservative compared to those estimated by the methods proposed in [8, 19–22]. In contrast, the current estimations are higher than those derived using the Nottingham-Schmertmann method [21] and the Dutch Method [22]. Specifically:

- The Meyerhof method [19] estimates pile capacity at approximately 150% of the ultimate load determined in this study.

Table 5 Tested uplift pile capacity compared to the estimated values of existing methods

Pile id	Ultimate uplift load capacity, Pu (kN)						
	Field	Method 1	Method 2	Method 3	Method 4	Method 5	Method 6
1	3029	4533	7252	2720	6346	2266	6044
2	956	1491	2386	895	2088	746	1988
3	3029	7068	11,309	4241	9896	3534	9425
4	1433	2092	3348	1255	2929	1046	2790
5	3711	5372	8595	3223	7521	2686	7163
6	1022	2166	3466	1300	3033	1083	2889
7	3711	5004	8007	3003	7006	2502	6673
8	3711	5513	8821	3308	7719	2757	7351
9	1523	3038	4862	1823	4254	1519	4051

Note Method 1: Meyerhof 1983; Method 2: Nottingham-Schmertmann 1975; Method 3: Schmertmann 1978; Method 4: Dutch Method 1976; Method 5: LCPL

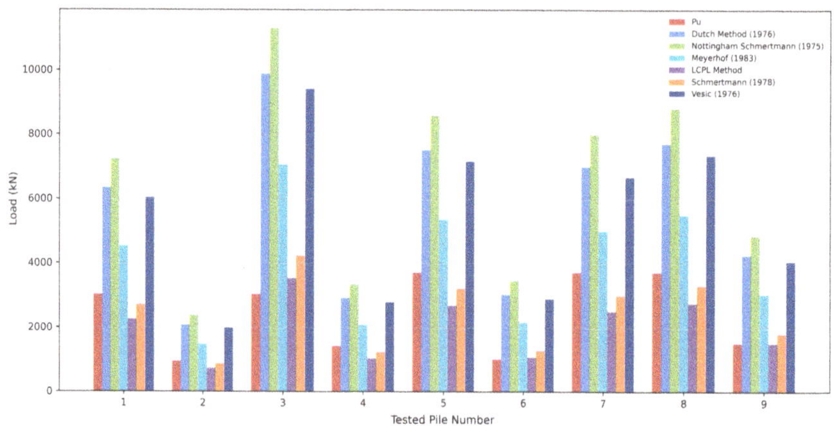

Fig. 5 Measured versus estimated ultimate load capacities (Pu) (kN) for the nine tested piles

- The methods proposed in [5, 8] estimate pile capacities exceeding 200% of the skin friction capacity calculated in this study.
- The methods proposed in [21, 22] estimate pile capacities ranging between 70 and 100% of the values predicted in this study.

In general, there is significant variability among scholars regarding pile skin friction estimates. Based on the load–displacement test graph, we conclude that our pile capacity tends to be conservative. To optimise the design, we propose increasing the skin friction factor $(ks) > 1$.

Furthermore, based on the results of 38 uplift pile load tests, the displacement corresponding to 200% of the design load remains within acceptable limits. This suggests that the ultimate unit capacity for drilled piles can be set at approximately $0.0045qc$, which aligns closely with Meyerhof's recommendations [19].

4 Discussion

The results presented in this study highlight the potential of machine learning techniques, particularly ensemble methods, for predicting the uplift behaviour of bored piles using derived CPT parameters. This section interprets these findings, discusses their implications and acknowledges the limitations of the current research.

4.1 Interpretation of Model Performance

The superior performance of the random forest (RF) and gradient boosting (GB) models, achieving near-perfect R2 values (0.999) according to the abstract, is a key finding. This significantly surpasses the performance of the simpler linear regression (LR), polynomial regression (PR) and even the multilayer perceptron (MLP) models. The success of RF and GB stems from their ensemble nature. These methods combine the predictions of numerous individual decision trees, enabling them to effectively capture complex, non-linear relationships and feature interactions within the dataset that simpler models might miss. Furthermore, ensemble methods are often more robust to noise and outliers, which could be particularly relevant given the use of augmented data derived from SPT-CPT correlations. The PR model showed moderate success (R2 = 0.87), suggesting some non-linearity in the relationship, while the MLP (R2 = 0.84) perhaps required further tuning or a different architecture to match the ensemble models. The basic LR model (R2 = 0.8) served as a baseline, confirming the complexity beyond a simple linear fit.

4.2   Significance of Feature Importance

The analysis revealed that pile perimeter (PP) and embedded length (L) were the dominant input features influencing uplift capacity predictions (Fig. 3). This aligns with the physical understanding that total shaft area, defined by PP $\times$ L, is a primary factor in skin friction capacity.

While fs (sleeve friction) and qc (tip resistance) are essential in defining unit shaft resistance, the variation in pile geometry had a stronger influence on this dataset. This is expected, as pile diameters varied significantly, and even small increases in diameter result in substantial increases in total surface area and uplift resistance.

The models detected these geometric effects more clearly than the smaller-scale variation in soil parameters. This suggests that when working with datasets involving piles of variable size, geometry may dominate model predictions unless soil properties vary substantially. This doesn't contradict geotechnical principles but reflects the relative scale of variation in the training data.

4.3   Validation Against Field Data and Existing Methods

The comparison of model predictions with the 28 full-scale field tests (summarised in Tables 3 and 4) provides crucial real-world validation. The high accuracy of the RF and GB models suggests they can effectively replicate observed pile behaviour under uplift loading, at least within the scope of the dataset used.

When comparing the study's results (field tests and ML predictions) against estimations from established CPT-based methods by other researchers (Fig. 5), the analysis reveals discrepancies. As suggested in the original document's conclusion, traditional methods might exhibit significant variability or conservatism. The ML models, trained directly on field performance data (albeit augmented and derived), potentially offer a more accurate or at least different perspective compared to generalised empirical formulas. The finding that a revised ultimate unit skin friction (e.g. $\approx 0.0045qc$) or an adjusted shaft coefficient (ks > 1) might better reflect the observed capacities suggests that the ML analysis, combined with field data, could lead to refinements in design approaches compared to potentially conservative standard practices.

4.4   Practical Implications and Limitations

The high predictive accuracy of the RF and GB models suggests their potential utility as tools for geotechnical engineers in predicting pile uplift displacement and capacity, complementing or potentially refining traditional methods. The derived equations from PR and LR, while less accurate, might offer simpler, more transparent (though less reliable) estimation tools.

However, several limitations must be acknowledged. Firstly, the study relies on a relatively small initial dataset (28 tests), necessitating data augmentation, which introduces synthetic data points that may not perfectly capture real-world variability. Secondly, the CPT parameters (qc, fs) were not all measured directly but were derived from SPT correlations. This introduces an inherent layer of uncertainty, as SPT-CPT correlations can be soil-type dependent and carry their error margins. The specific correlations used (qc≈0.4N60, Rf≈0.02) might not be universally applicable. Thirdly, the findings might be specific to the soil conditions and pile types represented in the original 28 tests.

5 Conclusion

This study investigated the application of various machine learning (ML) and artificial neural network (ANN) models—specifically linear regression (LR), polynomial regression (PR), multilayer perceptron (MLP), random forest (RF) and gradient boosting (GB)—for predicting the uplift load–displacement behaviour of bored piles using cone penetration test (CPT) parameters derived from standard penetration test (SPT) data and augmented through Gaussian noise injection.

The key findings are summarised as follows:

- Ensemble models, namely random forest (RF) and gradient boosting (GB), demonstrated significantly superior predictive performance (R2 ≈ 0.999) compared to MLP, PR and LR models in capturing the complex pile uplift behaviour based on the derived CPT data.
- Feature importance analysis confirmed that pile geometries as well as cone sleeve friction (fs) and cone tip resistance (qc) are the most influential parameters for predicting uplift capacity, aligning with geotechnical principles.
- The predictions from the best performing models (RF and GB) showed strong agreement when validated against the 28 full-scale field test results, indicating their potential to accurately replicate real-world pile performance within the context of this dataset.
- Comparison with existing CPT-based estimation methods suggested that some traditional approaches might be conservative, and the ML analysis combined with field data points towards potential refinements in design parameters, such as adjusting the shaft coefficient (ks > 1) or the ultimate unit skin friction (≈ 0.0045qc).

The primary contribution of this research lies in demonstrating the high efficacy of ensemble ML models (RF and GB) for pile uplift prediction, even when working with CPT data derived from correlations and augmented datasets. This highlights the potential for these advanced computational tools to supplement or refine traditional geotechnical design practices, offering potentially more accurate predictions based on performance data.

While the derived PR and LR equations offer simpler alternatives, their lower accuracy limits their reliability compared to the ensemble models. The practical implication is the potential deployment of validated RF or GB models as sophisticated tools for engineers assessing pile uplift capacity.

However, the study's limitations, including the reliance on derived CPT data, the relatively small initial dataset and potential site-specificity, must be considered. Future research should prioritise validating these findings using larger datasets incorporating direct CPT measurements across diverse geological settings and pile configurations. Further investigation into refining MLP models and applying these techniques to other pile loading conditions (compression, lateral) is also recommended.

References

1. Begemann R (1965) The friction jacket cone as an aid in determining the soil profile. Royal Dutch Society of Engineers, Amsterdam
2. Sanglerat S (1972) The penetrometer and soil exploration: interpretation of penetration diagrams—theory and practice. Elsevier, Amsterdam
3. Lunne T, Robertson PK, Powell JJM (1997) Cone penetration testing in geotechnical practice. Blackie Academic & Professional, London
4. Robertson PK (1990) Soil classification using the cone penetration test. Can Geotech J 27(1):151–158
5. Schmertmann JH (1978) Guidelines for cone penetration test: Performance and design. U.S. Department of Transportation, Federal Highway Administration, Report FHWA-TS-78209
6. Bustamante M, Gianeselli L (1982) Pile bearing capacity prediction by means of static penetrometer CPT. In: Proc 2nd European Symposium on Penetration Testing (ESOPT II), pp 493–500. Amsterdam
7. Eslami A, Fellenius BF (1997) Pile capacity by direct CPT and CPTu methods applied to 102 case histories. Can Geotech J 34(6):886–904
8. Vesic AS (1976) Design of pile foundations. Report No. FHWA-IF-71–59, U.S. Department of Transportation, Federal Highway Administration, Washington, D.C.
9. Ma C, Yu H-S (2002) A neural network approach to the determination of soil parameters from CPT data. Geotechnique 52(9):713–722
10. Nejad MP, Jaksa MB (2009) Application of artificial neural networks in the prediction of load–settlement behaviour of a pile. Comput Geotech 36(7):1246–1253
11. Nejad MP, Jaksa MB (2017) Prediction of axial load-settlement behavior of piles using artificial neural networks. Int J Geotech Eng 11(3):228–237
12. Shahin MA, Jaksa MB (2003) Pullout capacity prediction of small ground anchors using neural networks. J Comput Civ Eng 17(4):278–287
13. Shahin MA, Jaksa MB (2005) Neural networks for predicting pullout capacity of soil anchors. Adv Eng Softw 36(12):833–839
14. Shahin MA, Jaksa MB (2014) Using artificial neural networks for predicting the settlement of shallow foundations on granular soils. J South African Inst Civ Eng 56(1):36–44
15. Ardalan H, Eslami A (2009) Intelligent prediction of pile capacity using CPT data. Scientia Iranica 16(6):552–560
16. Kordjazi A, Shariatmadari A, Ghayoomi M (2014) Prediction of pile capacity using support vector machines. Arab J Sci Eng 39(7):5375–5386
17. Saeed M, Shahin M, Jaksa M (2014) Predicting pile capacity using hybrid intelligent computing techniques. Geotech Geol Eng 32(4):951–970

18. Bashar B (2017) Artificial neural networks for correlating CPT and SPT data. Eng Geol 229:1–11
19. Meyerhof GG (1983) Bearing capacity and settlement of pile foundations.J Geotech Eng, ASCE 109(3):195–228
20. Bustamante M, Gianeselli L (1982) Pile bearing capacity prediction by means of static penetrometer CPT. In: Proceedings of the 2nd European symposium on penetration testing, vol 2, pp 493–500. Balkema, Amsterdam, The Netherlands
21. Nottingham LC, Schmertmann JH (1975) Investigation of pile group interaction in sand. Report No. FHWA-RD-75–103, U.S. Department of Transportation, Federal Highway Administration, Washington, D.C
22. Dutch Foundation Manual (1976) Dutch method for foundation design based on cone penetration test. Royal Dutch Society of Engineers
23. Vesic AS (1973) Analysis of ultimate loads of shallow foundations. J Soil Mech Found Div 99(SM1):45–73

Evaluation of GCC Desert Sands as Potential Martian Regolith Simulants

W. Jrad and S. Alasadi

Abstract Replicating the physical and chemical properties of Martian regolith is essential for developing technologies in planetary exploration, rover mobility, and habitat construction. This study investigates desert sands from selected regions of the GCC, with samples collected from Saudi Arabia and Bahrain, to assess their suitability as terrestrial analogs for Martian regolith. Physical properties including grain size distribution, bulk density, and microscopic morphology were assessed alongside chemical and mineralogical compositions. Physical characterization was performed through grain size distribution analysis, bulk density measurement, and microscopic morphology observations. Chemical and mineralogical compositions were determined by different techniques. The results were compared with known Martian regolith properties obtained from Mars missions such as Curiosity and InSight. Preliminary findings suggest that certain sands from the Eastern Province of Saudi Arabia and central Bahrain exhibit particle size ranges, oxide content (SiO_2, Fe_2O_3, MgO), and mineral phases (quartz, hematite) closely matching Martian soil profiles. Mechanical behavior, including compressibility and shear strength, was also investigated to evaluate geotechnical analogies. This study demonstrates that Saudi and Bahraini desert sands present promising potential as natural, cost-effective simulants for use in Mars analog research facilities and experimental platforms.

Keywords Soil mechanics · Regolith simulant · Granular materials

W. Jrad (✉) · S. Alasadi
American University of Bahrain, Riffa, Bahrain
e-mail: wassim.jrad@aubh.edu.bh

S. Alasadi
e-mail: saraa.alasadi@aubh.edu.bh

© The Author(s) 2026

M. Alzaylaie et al. (eds.), *Geotechnical Innovation*, Lecture Notes in Civil Engineering 851, https://doi.org/10.1007/978-981-95-9215-9_6

1   Introduction

Understanding and replicating the physical and mechanical properties of Martian regolith is essential for advancing planetary science, particularly in the domains of robotic exploration, in situ resource utilization (ISRU), construction systems, and analog habitat development. Martian regolith, primarily composed of basaltic fines, angular particles, and iron-rich dust, presents unique challenges to surface operations. However, in situ testing on Mars remains highly limited, creating a strong need for Earth-based regolith simulants that accurately reproduce Martian surface conditions [1, 2].

Significant efforts have been made to analyze Martian soil properties using data from missions such as Viking, Phoenix, Curiosity, and InSight. These studies have provided crucial insights into regolith density, cohesion, bearing capacity, and particle morphology [3, 7, 8]. For instance, the InSight mission provided detailed pre-landing and post-landing regolith assessments, highlighting fine-grained, poorly compacted soils with friction angles typically ranging from 30° to 35° and minimal cohesion [2, 3]. The microscopic texture of Martian soil, characterized by sub-angular to angular particles, has also been documented through imaging technologies such as MAHLI and MECA [6].

To support Earth-based simulations, a variety of regolith simulants have been developed. These include synthetic materials like JSC Mars-1A and Mojave Mars Simulant (MMS), which aim to replicate the granulometry and mechanical behavior of Martian regolith [5]. Although widely used, these simulants can be expensive and limited in accessibility. Recent developments in large-volume lunar analogs such as EAC-1A also demonstrate the feasibility of designing accessible terrestrial substitutes for extraterrestrial soils [4, 9].

Additionally, new instrumentation, such as portable bevameters, is being explored to measure geotechnical behavior in analog environments, offering improved in situ validation capabilities for simulant performance [10]. These tools can assist in evaluating key parameters such as shear strength, compressibility, and surface traction across various simulant types.

This study builds upon the above research by investigating naturally occurring desert sands from Saudi Arabia (KSA) and Bahrain regions within the Gulf Cooperation Council (GCC) with environmental and geological similarities to Martian terrain. Through grain size analysis, microscopic imaging, and direct shear testing, this study aims to assess the suitability of these desert sands as cost-effective, accessible Martian regolith analogs for use in analog mission simulations and terrestrial testing facilities.

Fig. 1 Sand samples collected from the surface and placed in stainless-steel containers for analysis

2 Materials and Methods

2.1 Sampling Locations and Procedures

Two distinct desert regions in the Gulf Cooperation Council (GCC) were selected for this study based on their climatic conditions and potential geological similarities to Martian terrain:

- Eastern Province, Saudi Arabia, a zone dominated by extensive dune systems and weathered sands.
- Central Desert of Bahrain, a dry inland region featuring a mix of granular surface features shaped by natural processes.

At each location, sand samples were manually collected from the top 10 cm using standard scooping techniques and stainless-steel tools to minimize contamination (Fig. 1). Each sample was air-dried, sieved through a 2 mm mesh to remove oversized debris, and stored in sealed containers for laboratory analysis.

2.2 Grain Size Distribution

Grain size distribution is a critical parameter in evaluating the suitability of terrestrial materials as analogs for Martian regolith, as it influences soil behavior under mechanical stress, permeability, and interaction with rover wheels and excavation tools. A dry sieve analysis was performed on each sample using ASTM D6913 procedures. Mechanical shakers and a complete set of standard sieves ranging from 4.75 to 0.075 mm were used to determine particle size fractions (Fig. 2). The uniformity coefficient (Cu) and the coefficient of curvature (Cc) were calculated to assess the

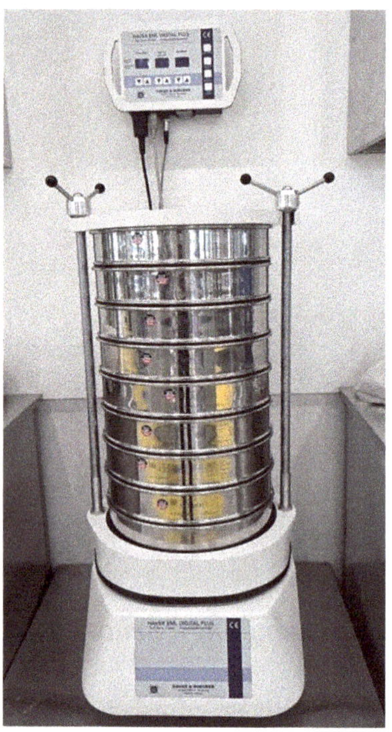

Fig. 2 Mechanical sieve shaker setup used

gradation characteristics of the soil samples, using the following standard equations:

$$Cu = \frac{D_{60}}{D_{10}} \tag{1}$$

$$Cc = \frac{D_{302}}{D_{10} \times D_{60}} \tag{2}$$

where D_{10}, D_{30}, and D_{60} represent the particle diameters corresponding to 10, 30 and 60% finer by weight, respectively, as determined from the grain size distribution curve.

These obtained values were compared with particle size ranges reported for Martian regolith by previous studies [1, 5] in the table.

2.3 Direct Shear Testing

To evaluate the mechanical behavior of the desert sands under simulated stress conditions, direct shear tests were conducted in accordance with ASTM D3080. A shear box apparatus (Fig. 3) was employed, with normal stress levels applied at 50, 100,

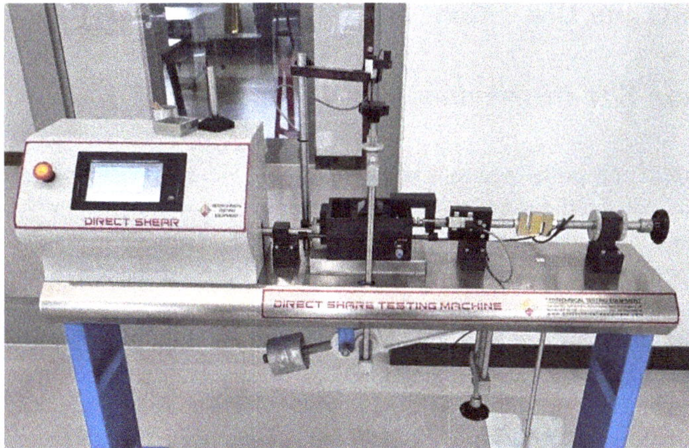

Fig. 3 Direct shear test apparatus used to evaluate the shear strength parameters of sand samples under controlled normal stress

and 150 kPa. The samples were compacted in three layers to replicate the estimated in situ densities. While tests were conducted at all three stress levels, only the results under 100 kPa are presented in this paper, as they provide a representative basis for evaluating the peak and residual shear responses. From these results, shear strength parameters were derived using the Mohr–Coulomb failure criterion, enabling the calculation of internal friction angle (ϕ) and cohesion (c).

2.4 Microscopic Observation

The morphology and angularity of the sand grains were examined using a stereo optical microscope under controlled lighting conditions. High-resolution images were captured for both Saudi and Bahraini samples. Observations focused on particle shape, surface texture, and presence of dust coatings, with comparisons drawn to microscopic imagery of Martian regolith from the Phoenix and Curiosity missions [6].

3   Results and Discussion

3.1   Grain Size Distribution

The results of the grain size analysis provide insight into the physical similarity between the collected desert sands and Martian regolith simulants. In what follows, the sieve analysis results are presented in Fig. 4 and compared to standard Martian regolith simulants in Table 1 to evaluate the suitability of the Saudi and Bahraini samples for analog applications.

Grain size analysis revealed that both the Saudi and Bahraini samples fall within the fine-to-medium sand category, aligning well with Martian regolith particle size estimates, which range from 50 to 500 μm [5, 6]. The Saudi sample exhibited a D50 of approximately 250 μm, while the Bahraini sample had a D50 of around 150 μm, suggesting slightly finer material.

Calculated parameters for the Saudi sample were Cu = 5.9, Cc = 0.76, indicating a well-graded soil. The Bahraini sample had Cu = 5.3, Cc = 0.67, consistent with a more uniform gradation. These values closely resemble those reported for synthetic

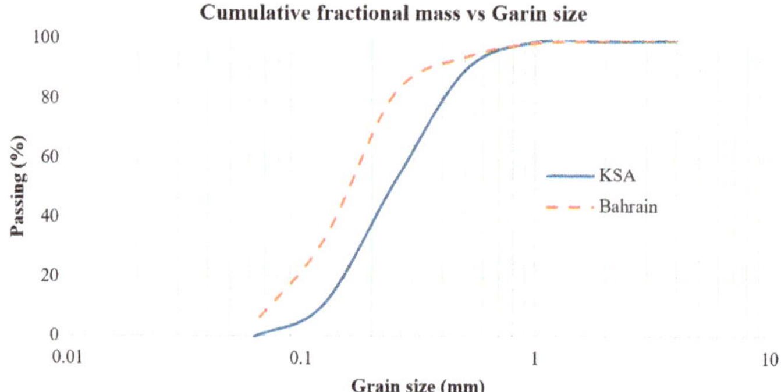

Fig. 4 Shows the grain size distribution curves for both samples, demonstrating a continuous distribution with minimal fines content, consistent with granular, cohesionless regolith simulants

Table 1 Comparison of grain size parameters for GCC samples and standard Martian regolith simulants [1]

Sample	Cu (Uniformity coefficient)	Cc (Curvature coefficient)	D50 (μm)
Bahrain	5.3	0.67	150
KSA	5.9	0.76	250
MSS-D [1]	3.27	0.44	100
Eifelsand [1]	3.36	1.04	310
Mojave [1]	5.58	0.72	640
MMS sand [5]	5.1	0.7	150

simulants such as Mojave Mars Simulant (MMS) and regolith inferred from InSight lander site studies [1, 2, 5]. These results indicate that both samples possess suitable gradation and particle size characteristics for analog simulation of Martian regolith.

3.2 Shear Strength Characteristics

The shear stress–displacement curves for both the Bahrain and Saudi samples (Fig. 5a and b) exhibit well-defined peaks followed by moderate softening, indicative of dense granular behavior. For the Bahrain sample, the peak shear stress reached 92 kPa, with a residual strength of 80 kPa. The Saudi sample showed a peak of 98 kPa and a residual strength of 85 kPa. These values fall within the range of Martian regolith simulants such as MSS-D and Mojave, previously reported in the literature [1, 5].

Based on the regression of multi-stress data, the internal friction angles were estimated as 33.9° for Bahrain and 35.2° for KSA. These results support the mechanical suitability of these GCC sands for regolith simulant applications. These values reflect dense granular behavior and are in agreement with the ranges reported for Martian regolith simulants such as Mojave and MSS-D [1, 5]. The results suggest that both materials possess sufficient internal friction to support analog applications involving mobility, traction, and load bearing under Martian gravity conditions. The Saudi sample demonstrated a slightly higher friction angle, possibly due to its coarser grading and slightly better packing during compaction.

These values are consistent with the shear strength ranges reported for Martian regolith simulants and inferred field data at the InSight and Viking landing sites, where values between 30° and 35° were observed [1, 3, 7]. The stress–displacement behavior also followed a linear trend without significant post-peak softening, which is expected in dry, sandy materials.

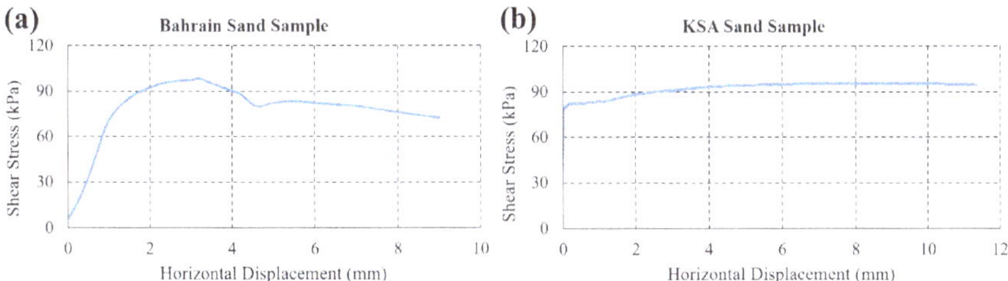

Fig. 5 **a**. Shear stress–displacement response of the Bahrain sand sample under 100 kPa normal load. **b**. Shear stress–displacement response of the Saudi sand sample under 100 kPa normal load

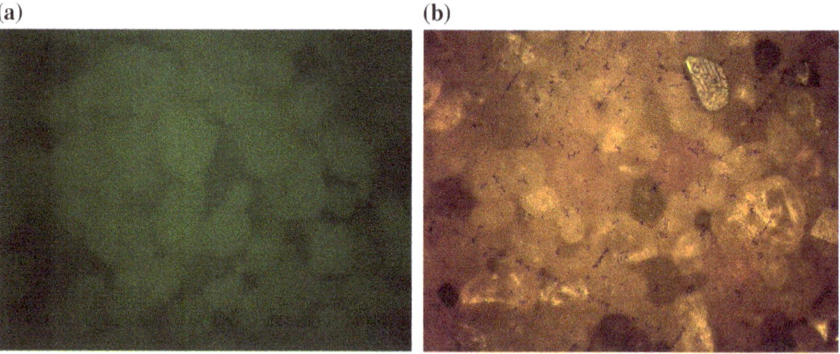

Fig. 6 a. Optical micrograph of the Bahraini sand. **b**. Optical micrograph of the KSA sand sample sample

3.3 Microscopic Observation

High-resolution images were captured for both Saudi and Bahraini samples and are presented in Fig. 6a and b, respectively.

The Bahraini sample (Fig. 6a) shows a mixture of sub-rounded to well-rounded grains with translucent mineral components, including notable polished surfaces indicative of aeolian weathering. A few grains also appear to exhibit internal striations or reflective inclusions, potentially feldspar or mica fragments.

In contrast, the Saudi sand sample (Fig. 6b) displays dominantly rounded particles with diffuse boundaries and a significant presence of fine dust coatings. This morphology is consistent with longer-term wind-blown processes and is similar to observations made of Martian aeolian deposits by Phoenix and Curiosity missions [6].

These textures and shapes are consistent with the morphology of Martian regolith particles described by Goetz et al. [6], which were identified as predominantly subangular to rounded grains. The visual similarity supports the potential of using these natural sands as cost-effective simulants for analog terrain construction and testing.

3.4 Comparative Relevance to Martian Regolith

The combined grain size, morphological, and shear strength properties suggest that GCC desert sands exhibit mechanical and physical behavior comparable to Martian regolith simulants used in engineering trials. While synthetic simulants offer precise geochemical tuning, the natural sands analyzed in this study provide a cost-effective and logistically practical alternative for experimental use in rover mobility testing,

Table 2 Major oxide compositions (wt.%) showing geochemical similarity between Bahrain, KSA, and Martian regolith

Oxide (wt.%)	Bahrain	KSA	Mars avg	Max from MER
SiO_2	45.12	47.75	46.52 ± 0.57	90.53
Fe_2O_3	4.73	4.75	4.2 ± 0.54	18.42
MgO	8.12	9.31	8.93 ± 0.45	16.46

analog mission setups, and small-scale ISRU prototyping. Additional chemical analysis, including oxide composition, is ongoing, and preliminary findings show encouraging similarity with the benchmark Martian soil data [5], particularly in SiO_2, Fe_2O_3, and MgO levels.

Table 2 demonstrates that the oxide composition of Bahrain and KSA sands is consistent with Martian soil characteristics reported by missions such as Curiosity and MER, reinforcing their suitability as natural analog materials. Further geochemical and mineralogical analyses will be presented in future publications.

4   Conclusion

This study evaluated the geotechnical behavior of desert sands from Saudi Arabia and Bahrain as potential natural analogs for Martian regolith. The samples demonstrated grain size distributions, shear strength parameters, and morphological features consistent with recognized Martian soil simulants, such as Mojave and MSS-D. Shear tests yielded peak strengths of 90–100 kPa and friction angles of 33.9°–35.2°, aligning closely with reported Martian surface soil values.

Preliminary chemical analyses revealed oxide compositions particularly in SiO_2, Fe_2O_3, and MgO that are comparable to Martian regolith values reported by the MER and Curiosity missions. These findings enhance the case for using naturally available GCC desert sands in Mars analog research, offering a cost-effective and accessible alternative to synthetic simulants.

This paper presents a subset of a broader ongoing investigation. Future work will expand on these results through detailed X-ray diffraction (XRD), X-ray fluorescence (XRF), and mineralogical characterization, along with advanced geotechnical testing. The goal is to develop a robust, regionally sourced simulant database to support planetary exploration research and engineering simulation across the Middle East.

References

1. Delage P, Karakostas F, Dhemaied A et al (2017) An investigation of the mechanical properties of some Martian regolith simulants with respect to the surface properties at the InSight mission landing site. Space Sci Rev 211:1–45. https://doi.org/10.1007/s11214-017-0339-7

2. Morgan P, Grott M, Knapmeyer-Endrun B et al (2018) A pre-landing assessment of regolith properties at the InSight landing site. Space Sci Rev 214:104. https://doi.org/10.1007/s11214-018-0537-y

3. Golombek M, Grott M, Kargl G et al (2018) Geology and physical properties investigations by the InSight lander. Space Sci Rev 214:84. https://doi.org/10.1007/s11214-018-0512-7

4. Engelschiøn VS, Eriksson SR, Cowley A et al (2020) EAC-1A: A novel large-volume lunar regolith simulant. Sci Rep 10:5473. https://doi.org/10.1038/s41598-020-62312-4

5. Peters GH, Carey JP, Anderson AC et al (2008) Mojave Mars simulant: A new Martian soil simulant. Icarus 197:470–479. https://doi.org/10.1016/j.icarus.2008.05.004

6. Goetz W, Hviid SF, Madsen MB et al (2010) Microscopy analysis of soils at the Phoenix landing site, Mars: Classification of soil particles and description of their optical and magnetic properties. J Geophys Res Planets 115:E00E22. https://doi.org/10.1029/2009JE003437

7. Golombek M, Warner N, Grant JA et al (2017) Selection of the InSight landing site. Space Sci Rev 211:5–95

8. Delage P, Marteau E, Vrettos C, Golombek M, Ansan V, Banerdt WB, Williams R (2022) The mechanical properties of the Martian soil at the InSight landing site. In: Proceeedings 20th international conference on soil mechanics and geotechnical engineering, Sydney

9. Lucas MP, Neal CR, Long-Fox JM, Britt D (2024) Simulating lunar highlands regolith profiles on Earth to inform infrastructure development and ISRU activities on the Moon. Acta Astronaut. https://www.sciencedirect.com/science/article/pii/S0094576524004557

10. Saito J, Wang H, Zhu F (2024) Portable bevameter design for geotechnical characterization on planetary surfaces. In: 2024 international conference on Aerospace and Space Technology (ICAST). IEEE. https://ieeexplore.ieee.org/document/10687506

GPSR Compliance
The European Union's (EU) General Product Safety Regulation (GPSR) is a set
of rules that requires consumer products to be safe and our obligations to
ensure this.

If you have any concerns about our products, you can contact us on

ProductSafety@springernature.com

In case Publisher is established outside the EU, the EU authorized
representative is:

Springer Nature Customer Service Center GmbH
Europaplatz 3
69115 Heidelberg, Germany

www.ingramcontent.com/pod-product-compliance
Ingram Content Group UK Ltd.
Pitfield, Milton Keynes, MK11 3LW, UK
UKHW021009080726
473054UK00003B/16